SpringerBriefs in Applied Sciences and Technology

SpringerBriefs present concise summaries of cutting-edge research and practical applications across a wide spectrum of fields. Featuring compact volumes of 50 to 125 pages, the series covers a range of content from professional to academic.

Typical publications can be:

- A timely report of state-of-the art methods
- An introduction to or a manual for the application of mathematical or computer techniques
- A bridge between new research results, as published in journal articles
- A snapshot of a hot or emerging topic
- An in-depth case study
- A presentation of core concepts that students must understand in order to make independent contributions

SpringerBriefs are characterized by fast, global electronic dissemination, standard publishing contracts, standardized manuscript preparation and formatting guidelines, and expedited production schedules.

On the one hand, **SpringerBriefs in Applied Sciences and Technology** are devoted to the publication of fundamentals and applications within the different classical engineering disciplines as well as in interdisciplinary fields that recently emerged between these areas. On the other hand, as the boundary separating fundamental research and applied technology is more and more dissolving, this series is particularly open to trans-disciplinary topics between fundamental science and engineering.

Indexed by EI-Compendex, SCOPUS and Springerlink.

Tin-Chih Toly Chen

Neutrosophic Collaborative Intelligence and Systems

Methodology, System Architecture, and Applications in Engineering

Springer

Tin-Chih Toly Chen
Department of Industrial Engineering
and Management
National Yang Ming Chiao Tung University
Hsinchu, Taiwan

ISSN 2191-530X ISSN 2191-5318 (electronic)
SpringerBriefs in Applied Sciences and Technology
ISBN 978-3-032-34266-9 ISBN 978-3-032-34267-6 (eBook)
https://doi.org/10.1007/978-3-032-34267-6

This Springer imprint is published by the registered company Springer Nature Switzerland AG
The registered company address is: Gewerbestrasse 11, 6330 Cham, Switzerland

To the pioneers of systems engineering, collaborative intelligence, and fuzzy logic—whose foundational work continually inspires innovative leaps into the realm of uncertainty and indeterminacy. May this framework empower future engineers and decision makers to embrace complexity rather than shy away from it.

Foreword

In an era of relentless technological advancement, dealing with indeterminacy has become one of the most critical challenges in quantitative and decision sciences. While traditional models are forced to prematurely discard indeterminacy to reach a consensus, the concept of neutrosophic collaborative intelligence (NCI) revolutionizes this process. Evolving from fuzzy collaborative intelligence (FCI), NCI offers a sophisticated mechanism to process expert judgments by retaining indeterminacy throughout the reasoning process.

This book elegantly bridges the gap between theoretical methodology and practical application. It systematically introduces the system architecture and mathematical foundations of NCI approaches, demonstrating its prowess in multi-criteria decision-making. Whether you are a researcher expanding the frontiers of logic, or a practicing engineer and manager striving to optimize operations amid indeterminate parameters, this text provides an indispensable lens through which complex, real-world problems can be solved.

Hsinchu, Taiwan

Tin-Chih Toly Chen

Acknowledgments The completion of this book would not have been possible without the invaluable insights, continuous support, and collective wisdom of many individuals. I would like to express my deepest gratitude to my academic colleagues, industry colleagues, and the broader research community in operations management, mechanical engineering, and information management who contributed to the iterative development of this framework.

Special thanks go to the editorial and production teams for their meticulous review and encouragement. Finally, to my family and friends, thank you for your endless patience, unwavering support, and continuous inspiration during the writing of this book.

Competing Interests The author has no competing interests to declare that are relevant to the content of this manuscript.

Contents

Chapter 1
Neutrosophic Collaborative Intelligence

Abstract Neutrosophic collaborative intelligence (NCI) evolved from fuzzy collaborative intelligence (FCI). Therefore, this chapter first introduces FCI and fuzzy collaborative systems (FCSs), and shows that FCSs are distinct from existing fuzzy ensembles, fuzzy inference systems, fuzzy weighted averaging, fuzzy geometric mean, etc. NCI systems are also expected to differ from existing neutrosophic ensembles, neutrosophic inference systems, and neutrosophic weighted averaging. Subsequently, neutrosophic sets and operations are described, followed by the formal introduction of NCI and neutrosophic collaborative systems (NCSs). Existing neutrosophic collaborative intelligence and systems are then categorized along four dimensions. An implementation process for NCI and NCS is also provided. The chapter then provides examples of various types of neutrosophic collaborative intelligence and systems. Fields in which NCI methods are most frequently applied include neutrosophic collaborative decision-making, neutrosophic collaborative evaluation, neutrosophic collaborative management, and neutrosophic collaborative learning. In contrast, neutrosophic collaborative classification, neutrosophic collaborative recommendation, and neutrosophic collaborative forecasting are fields where NCI methods are rarely applied.

Keyword Fuzzy collaborative intelligence · Fuzzy collaborative system · Neutrosophic set · Neutrosophic collaborative intelligence · Neutrosophic collaborative system

1.1 From Fuzzy Collaborative Intelligence to Neutrosophic Collaborative Intelligence

Neutrosophic collaborative intelligence (NCI) evolved from fuzzy collaborative intelligence (FCI) [1]. Another evolution of FCI leads to fuzzy federated learning or fuzzy federated intelligence [2] (see Fig. 1.1). The former is grounded in parallel and distributed computing, while the latter considers or attempts to simulate human

T.-C.T. Chen, *Neutrosophic Collaborative Intelligence and Systems*,
SpringerBriefs in Applied Sciences and Technology,
https://doi.org/10.1007/978-3-032-34267-6_1

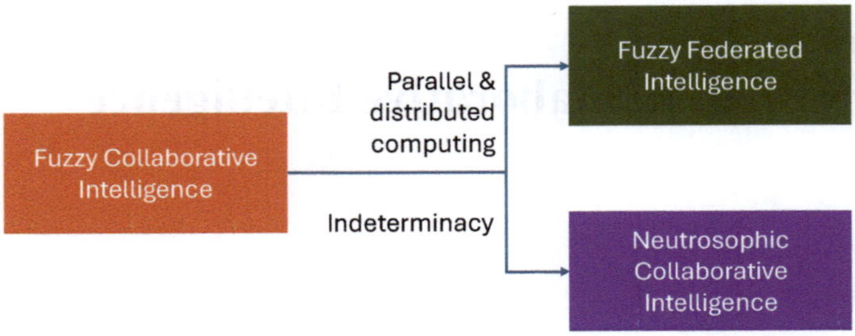

Fig. 1.1 Evolutions of FCI

indeterminacy [3]. Therefore, this chapter first reviews the methodologies, system architectures, and applications of FCI as follows.

FCI aims to conduct multiple fuzzy analyses of a problem from diverse perspectives to raise the chance that no relevant aspects of the problem are ignored. FCI is especially critical as Internet applications are becoming more widespread. Under such environments, dealing with disparate data sources is becoming increasingly popular. Technical constraints, security issues, and privacy considerations often limit access to some sources. Therefore, the concepts of FCI and collaborative fuzzy modeling have been proposed, and certain so-called fuzzy collaborative systems (FCSs) are being established [4–6]. So far, FCI has become one of the important trends in the field of collective intelligence [7, 8].

In a FCS, some experts, agents, or systems with various backgrounds, motives, or perspectives are trying to achieve a common target. Since these experts, agents, or systems have different knowledge and points of view, they may use various methods to model, identify, or control the common target. Specifically, fuzzy methods are used to account for uncertainties in inputs, outputs, expert knowledge, and/or viewpoints. The key to such a system is that these experts, agents, or systems share and exchange their observations, settings, experiences, and knowledge of each other when achieving the common goal. This features a FCS distinct from an ensemble of multiple fuzzy systems. To account for all viewpoints, in FCSs, fuzzy outputs typically include actual values, which differs from fuzzy inference systems (FISs). Furthermore, in FCSs used for decision-making, fuzzy intersection (FI) [9], rather than fuzzy weighted average (FWA), is employed to aggregate expert judgments or evaluations, thereby including actual values (reasonable decisions) in the aggregation results [10]. In this way, the forecasting or decision-making precision is enhanced.

Figure 1.2 shows the properties of FCSs distinct from existing fuzzy ensembles, FISs, and FWA. NCI systems are also expected to differ from existing neutrosophic ensembles, neutrosophic inference systems, and neutrosophic weighted averaging.

Although there has been some literature about fuzzy collaborative intelligence and systems, considerable room for development still exists in this field [6], for example.

- how to fuzzify a method or system to meet the requirements of a fuzzy collaborative intelligence method or system,

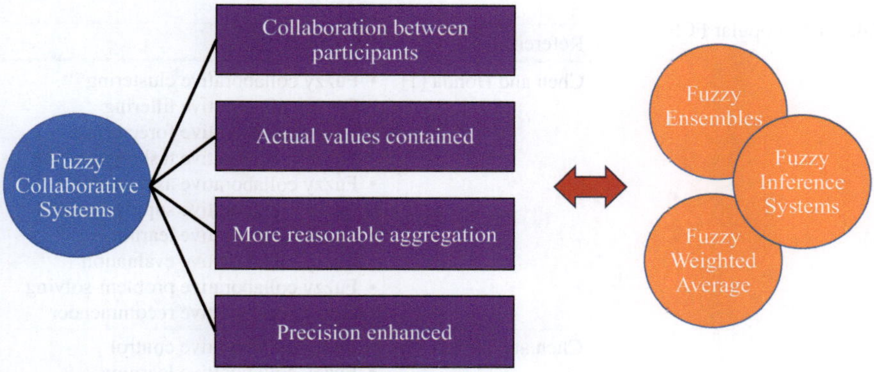

Fig. 1.2 Properties of FCSs distinct from existing fuzzy ensembles, FISs, and FWA

- how to facilitate the collaboration among experts, agents, or systems,
- how to aggregate the views of, and the results by, experts, and
- whether individual fuzzy predictions are equally accurate, or how to distinguish between different accuracy levels of individual fuzzy forecasts.

Another major research direction is to apply various types of fuzzy numbers to build fuzzy collaborative systems. So far, most existing fuzzy collaboration systems have been used for control [11, 12] (due to the popularity of robots), learning [13, 14], and evaluation [15, 16].

Both Chen and Honda [1] and Chen and Honda [6] conducted statistics on the popularity of various FCSs. The most popular FCSs are summarized in Table 1.1. In the past, fuzzy collaborative clustering attracted attention due to distributed databases and applications, but nowadays, with the booming development of autonomous vehicles and robots, fuzzy collaborative control has become a research hotspot.

1.2 Neutrosophic Sets and Operations

A neutrosophic set is defined as follows.

Definition 1 [17]

Let **X** be the data space, $x \in \mathbf{X}$. A neutrosophic set $\tilde{A}$ in **X** has three membership functions: the truth membership function $T_{\tilde{A}}(x)$, the indeterminacy membership function $I_{\tilde{A}}(x)$, and the falsity membership function $F_{\tilde{A}}(x)$. $T_{\tilde{A}}(x), I_{\tilde{A}}(x)$, and $F_{\tilde{A}}(x)$ are real standard or nonstandard subsets of]-0, $1+[$. $0^- \le T_{\tilde{A}}(x) - I_{\tilde{A}}(x) + F_{\tilde{A}}(x) \le 3^+$.

Definition 2 [18]

Table 1.1 Popular FCSs

Reference	Popular FCSs
Chen and Honda [1]	• Fuzzy collaborative clustering • Fuzzy collaborative filtering • Fuzzy collaborative forecasting • Fuzzy collaborative task assignment • Fuzzy collaborative assessment • Fuzzy collaborative supply chain • Fuzzy collaborative learning • Fuzzy collaborative evaluation • Fuzzy collaborative problem-solving • Fuzzy collaborative recommender
Chen and Honda [6]	• Fuzzy collaborative control • Fuzzy collaborative learning • Fuzzy collaborative evaluation • Fuzzy collaborative assessment • Fuzzy collaborative problem-solving • Fuzzy collaborative clustering • Fuzzy collaborative supply chain • Fuzzy collaborative forecasting • Fuzzy collaborative filtering • Fuzzy collaborative recommender

A triangular neutrosophic number (TNN) $\tilde{A} = (A_1, A_2, A_3); \alpha_{\tilde{A}}, \theta_{\tilde{A}}, \beta_{\tilde{A}}$ is a special neutrosophic set in $\mathbf{R}$ with the following truth, indeterminacy, and falsity membership functions:

$$T_{\tilde{A}}(x) = \begin{cases} \alpha_{\tilde{A}}\left(\frac{x-A_1}{A_2-A_1}\right) & if \quad A_1 \le x < A_2 \\ \alpha_{\tilde{A}}\left(\frac{A_3-x}{A_3-A_2}\right) & if \quad A_2 \le x < A_3 \\ 0 & otherwise \end{cases} \tag{1.1}$$

$$I_{\tilde{A}}(x) = \begin{cases} \frac{A_2-x+\theta_{\tilde{A}}(x-A_1)}{A_2-A_1} & if \quad A_1 \le x < A_2 \\ \frac{x-A_2+\theta_{\tilde{A}}(A_3-x)}{A_3-A_2} & if \quad A_2 \le x < A_3 \\ 0 & otherwise \end{cases} \tag{1.2}$$

$$F_{\tilde{A}}(x) = \begin{cases} \frac{A_2-x+\beta_{\tilde{A}}(x-A_1)}{A_2-A_1} & if \quad A_1 \le x < A_2 \\ \frac{x-A_2+\beta_{\tilde{A}}(A_3-x)}{A_3-A_2} & if \quad A_2 \le x < A_3 \\ 0 & otherwise \end{cases} \tag{1.3}$$

where $\alpha_{\tilde{A}}, \theta_{\tilde{A}}$, and $\beta_{\tilde{A}}$ indicate the highest truth membership degree, the lowest indeterminacy membership degree, and the lowest falsity membership degree, respectively. An example is given in Fig. 1.3. There are other types of neutrosophic numbers, such as trapezoidal neutrosophic numbers [19], neutrosophic fuzzy hypersoft numbers [20], bipolar neutrosophic numbers [21], and single-valued neutrosophic numbers [22].

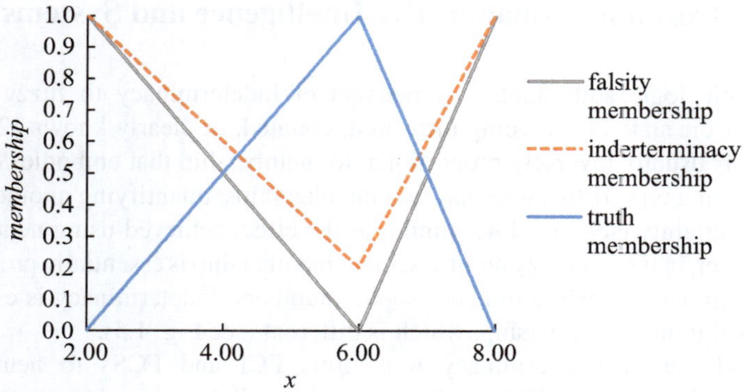

Fig. 1.3 Triangular neutrosophic number [23]

Some arithmetic operations on TNNs are defined as follows, using product t-norm and t-conorm [24].

Theorem 1 [24]

- Addition: $\tilde{A}(+)\tilde{B} = (A_1 + B_1, A_2 + B_2, A_3 + B_3); \alpha_{\tilde{A}} + \alpha_{\tilde{B}} - \alpha_{\tilde{A}}\alpha_{\tilde{B}}, \theta_{\tilde{A}}\theta_{\tilde{B}}, \beta_{\tilde{A}}\beta_{\tilde{B}}$
- Subtraction: $\tilde{A}(-)\tilde{B} = (A_1 - B_3, A_2 - B_2, A_3 - B_1); \alpha_{\tilde{A}} + \alpha_{\tilde{B}} - \alpha_{\tilde{A}}\alpha_{\tilde{B}}, \theta_{\tilde{A}}\theta_{\tilde{B}}, \beta_{\tilde{A}}\beta_{\tilde{B}}$
- Inverse: $\frac{1}{\tilde{A}} \cong \left(\frac{1}{A_3}, \frac{1}{A_2}, \frac{1}{A_1}\right); \alpha_{\tilde{A}}, \theta_{\tilde{A}}, \beta_{\tilde{A}}$ if $A_1 \geq 0$
- Multiplication: $\tilde{A}(\times)\tilde{B} \cong (A_1B_1, A_2B_2, A_3B_3); \alpha_{\tilde{A}}\alpha_{\tilde{B}}, \theta_{\tilde{A}} + \theta_{\tilde{B}} - \theta_{\tilde{A}}\theta_{\tilde{B}}, \beta_{\tilde{A}} + \beta_{\tilde{B}} - \beta_{\tilde{A}}\beta_{\tilde{B}}$ if $A_1, B_1 \geq 0$
- Division: $\tilde{A}(/)\tilde{B} \cong (A_1/B_3, A_2/B_2, A_3/B_1); \alpha_{\tilde{A}}\alpha_{\tilde{B}}, \theta_{\tilde{A}} + \theta_{\tilde{B}} - \theta_{\tilde{A}}\theta_{\tilde{B}}, \beta_{\tilde{A}} + \beta_{\tilde{B}} - \beta_{\tilde{A}}\beta_{\tilde{B}}$ if $B_1 \geq 0$
- Square root: $\sqrt{\tilde{A}} \cong \left(\sqrt{A_1}, \sqrt{A_2}, \sqrt{A_3}\right); \alpha_{\tilde{A}}, \theta_{\tilde{A}}, \beta_{\tilde{A}}$ if $A_1 \geq 0$

which can effectively avoid the drawback of the traditional method based on intersection (minimum) and union (maximum) operators as follows.

Theorem 2 [25]

- Addition: $\tilde{A}(+)\tilde{B} = (A_1 + B_1, A_2 + B_2, A_3 + B_3); \alpha_{\tilde{A}} \wedge \alpha_{\tilde{B}}, \theta_{\tilde{A}} \vee \theta_{\tilde{B}}, \beta_{\tilde{A}} \vee \beta_{\tilde{B}}$
- Subtraction: $\tilde{A}(-)\tilde{B} = (A_1 - B_3, A_2 - B_2, A_3 - B_1); \alpha_{\tilde{A}} \wedge \alpha_{\tilde{B}}, \theta_{\tilde{A}} \vee \theta_{\tilde{B}}, \beta_{\tilde{A}} \vee \beta_{\tilde{B}}$
- Inverse: $\frac{1}{\tilde{A}} \cong \left(\frac{1}{A_3}, \frac{1}{A_2}, \frac{1}{A_1}\right); \alpha_{\tilde{A}}, \theta_{\tilde{A}}, \beta_{\tilde{A}}$ if $A_1 \geq 0$
- Multiplication: $\tilde{A}(\times)\tilde{B} \cong (A_1B_1, A_2B_2, A_3B_3); \alpha_{\tilde{A}} \wedge \alpha_{\tilde{B}}, \theta_{\tilde{A}} \vee \theta_{\tilde{B}}, \beta_{\tilde{A}} \vee \beta_{\tilde{B}}$ if $A_1, B_1 \geq 0$
- Division: $\tilde{A}(/)\tilde{B} \cong (A_1/B_3, A_2/B_2, A_3/B_1); \alpha_{\tilde{A}} \wedge \alpha_{\tilde{B}}, \theta_{\tilde{A}} \vee \theta_{\tilde{B}}, \beta_{\tilde{A}} \vee \beta_{\tilde{B}}$ if $B_1 \geq 0$
- Square root: $\sqrt{\tilde{A}} \cong \left(\sqrt{A_1}, \sqrt{A_2}, \sqrt{A_3}\right); \alpha_{\tilde{A}}, \theta_{\tilde{A}}, \beta_{\tilde{A}}$ if $A_1 \geq 0$

that is, the judgment with the highest (or lowest) membership dominates the derivation results.

1.3 Neutrosophic Collaborative Intelligence and Systems

Neutrosophic logic adds another dimension of indeterminacy to fuzzy logic to account for the state of not being measured, counted, or clearly known [26]. Indeterminacy is usually inversely proportional to membership that embodies certainty. Each value in a type-II fuzzy set has two memberships, quantifying another type of human uncertainty (see Fig. 1.4), similar to the effect achieved using neutrosophic sets. However, in type-II fuzzy numbers, inner membership is essentially proportional to outer membership, while in neutrosophic numbers, indeterminacy is essentially proportional to non-membership, which is different (see Fig. 1.5).

The addition of indeterminacy transforms FCI and FCSs to neutrosophic collaborative intelligence (NCI) and neutrosophic collaborative systems (NCSs).

Similar to the classification of FCI and FCSs, neutrosophic collaborative intelligence (NCI) and neutrosophic collaborative systems (NCSs) can be classified along four dimensions (see Fig. 1.6):

- What is the type (or function) of neutrosophic collaborative intelligence and systems?

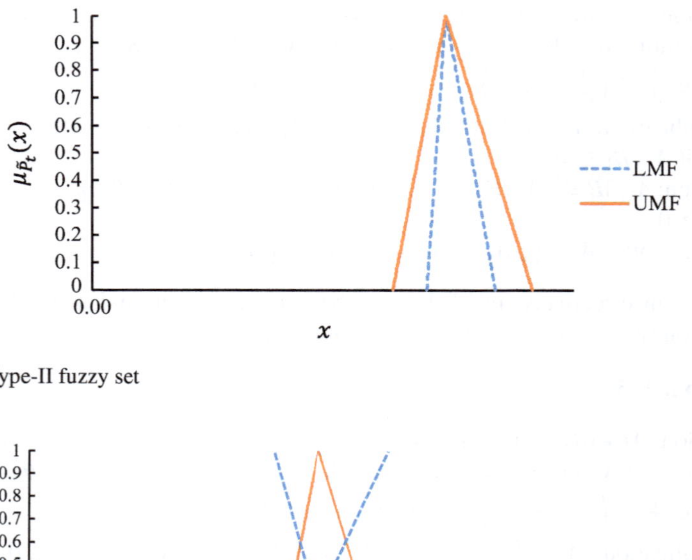

Fig. 1.4 Type-II fuzzy set

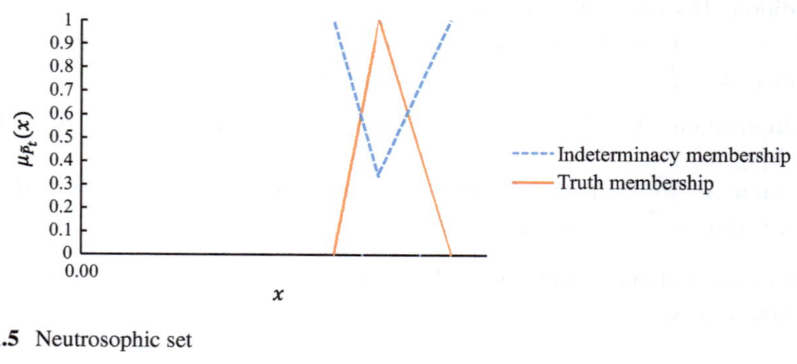

Fig. 1.5 Neutrosophic set

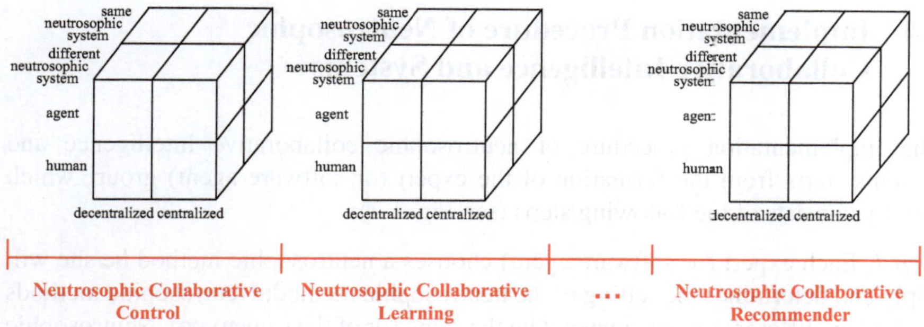

Fig. 1.6 Four-dimensional classification of neutrosophic collaborative intelligence and systems

- Is data access decentralized or centralized?
- Are real experts or software agents (virtual experts) used: In a NCI approach, domain experts must be consulted to ensure that the results produced meet their requirements. However, if domain experts are unavailable for consultation, a "virtual expert" committee can be formed to provide a set of opinions for the NCI approach. This does not mean that the opinion set can be randomly generated, as an unrealistic opinion set may result in no feasible solution to the optimization problem or increase the difficulty of finding the optimal solution. However, the concept of indeterminacy may not be easy to comprehend by agents, which makes agent-based NCI and NCSs distinct from agent-based FCI and FCSs.
- The similarities or differences between the neutrosophic methods (or systems) used on different locations (or by different experts).
- How does indeterminacy change when multiple experts, agents, or systems collaborate using neutrosophic methods?
- Is the function used for deriving indeterminacy during collaboration the same as that for deriving membership?
- Is there a single function for such derivation: So far, more than one type of function has been commonly applied to aggregate the results by multiple experts, agents, or systems using neutrosophic methods [26, 27].

Specifically, when the neutrosophic methods used by experts (or agents) are of the same type, the NCI approach is called a homogeneous NCI approach; otherwise, the NCI approach will be a heterogeneous NCI approach.

1.4 Implementation Procedure of Neutrosophic Collaborative Intelligence and Systems

The implementation procedure of neutrosophic collaborative intelligence and systems starts from the formation of the expert (or software agent) group, which usually consists of the following steps (see Fig. 1.7):

Step 1. Each expert (or software agent) chooses a neutrosophic method he/she will apply and determines the setting of the neutrosophic method: Neutrosophic methods applied by different experts may not be the same (or of the same type). Neutrosophic methods are homogeneous if they are of the same type. When agents are used, the settings will be casual and less practical.

Step 2. Each expert applies his/her neutrosophic method to fulfill the task: Inputs to a neutrosophic method may need to be transformed into neutrosophic sets [25], and then neutrosophic arithmetic operations are performed on them, which generate neutrosophic outputs. Among these, transforming inputs to neutrosophic sets, i.e.,

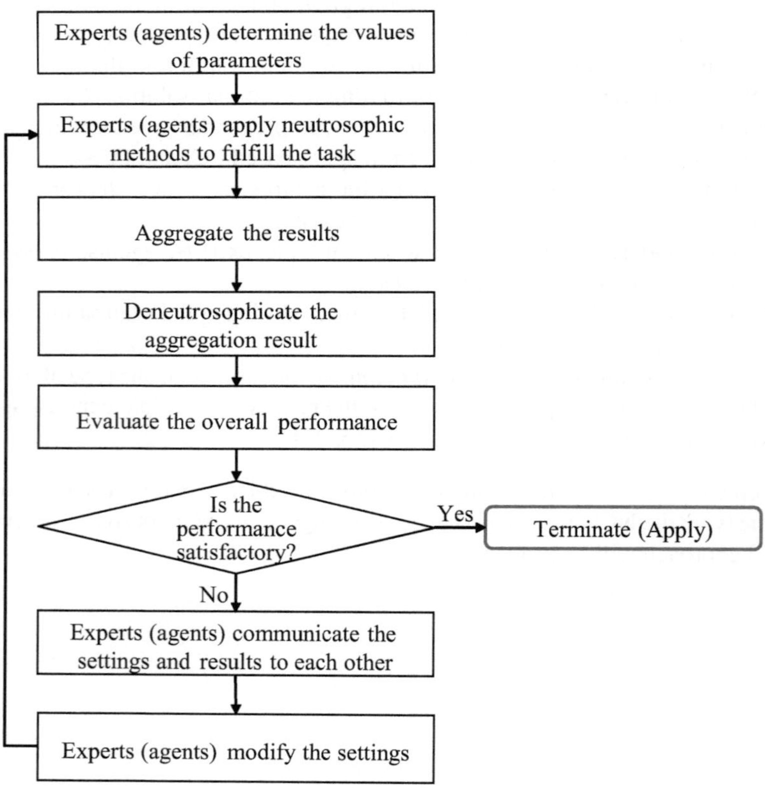

Fig. 1.7 Implementation procedure of neutrosophic collaborative intelligence and systems

neutrosophication, is a critical yet challenging issue. For example, in neutrosophic decision-making problems, the relative priorities of criteria are given in neutrosophic sets, the indeterminacy of which is pre-specified, which may be groundless. In neutrosophic time-series forecasting problems, the indeterminacy of inputs may be assigned by domain experts subjectively or derived from the short-term upward or downward trend in the time series, which is interesting but may be arbitrary.

Step 3. Each expert communicates his/her view and results to other experts, perhaps with the aid of a centralized collaboration mechanism.

Step 4. Upon receipt, the expert may choose to adjust his/her setting and re-apply the neutrosophic method to fulfill the task.

Step 5. The results by all experts are aggregated.

Step 6. The aggregation result is deneutrosophicated to generate a single, representative value, which is usually achieved using a score function: For example, the score function value of a triangular neutrosophic number (TNN) $\tilde{A} = A_1, A_2, A_3; \alpha_{\tilde{A}}, \theta_{\tilde{A}}, \beta_{\tilde{A}}$ can be evaluated by [20]

$$S(\tilde{A}) = \frac{1}{8}(A_1 + A_2 + A_3) \cdot \left(2 + \alpha_{\tilde{A}} - \theta_{\tilde{A}} - \beta_{\tilde{A}}\right) \tag{1.4}$$

Just like there are many defuzzification methods, other score functions are also applicable.

Step 7. The overall performance is evaluated.

Step 8. The collaboration process is terminated if the improvement in the overall performance becomes negligible. Otherwise, return to Step 3.

1.5 Applications of Neutrosophic Collaborative Intelligence and Systems

Some applications of NCI and NCSs are reviewed as follows.

Wafer foundries face cultural differences when building localized wafer fabs, so they have proposed multiple measures to mitigate the impact of cultural differences. However, given the limited budget, time and other resources, it is critical to prioritizing these measures. Due to the lack of experience and references, how to prioritize criteria and evaluate the performances of measures remain indeterminate. To address this issue, Chiu and Chen [23] proposed a NCI approach, in which each expert used the calibrated neutrosophic geometric mean (cNGM) to derive the neutrosophic weights of criteria for prioritizing mitigating measures. The neutrosophic technique for order of preference by similarity to the ideal solution (NTOPSIS) was then used to evaluate the suitability of each measure. Finally, the neutrosophic weighted intersection (NWI) operator was devised to aggregate the evaluation results by all experts.

A systematic procedure was also established to prioritize these measures. The NCI approach has been applied to a real case. According to the experimental results, legal resolvability was emphasized as the most decisive factor behind why some mitigating measures were more favorable than others. The most suitable measure for mitigating the impact of cultural differences was accelerating yield learning through early mass production, while the least suitable measure was designing a staggered shift system. Other conclusions were indeterminate because experts' more uncertain judgments had a greater impact on the two measures to be compared than their more certain judgments.

Neutrosophic group decision-making methods were proposed to better capture indeterminacy inherent in expert judgments [28]. Ihsan et al. [28] proposed the concept of fuzzy parameterized single-valued neutrosophic soft sets, the complement of such a sets, and the union and intersection of such sets. Their methodology was applied to a group multi-criteria decision-making (MCDM) problem, in which several experts were asked to evaluate and compare the suitabilities of several workstation computers for a business by considering various factors such as buffer memory, access time, platter diameter, spindle speed, drive form factor, and disk partitioning.

Basset et al. [25] combined neutrosophic analytic hierarchy process (AHP) and strengths, weaknesses, opportunities, and threats (SWOT) analysis to compare a set of alternative strategies for the Starbucks company developed on the basis of factors and subfactors identified using a SWOT analysis. The pairwise comparison results of the relative priorities of criteria were given in TNNs and converted into their score function values. Then, the priorities of criteria were derived by arithmetic averaging, i.e., the priority of a criterion was proportional to the sum of the corresponding row in the judgment matrix.

Junaid et al. [29] established a neutrosophic AHP and TOPSIS framework for supply chain risk assessment in the automotive industry of Pakistan. The relative priorities of criteria were given in TNNs. Similar to Basset et al. [26], pairwise comparison results were transformed into their score function values, so that arithmetic averaging could be easily applied to derive the priorities of criteria. Therefore, crisp TOPSIS was applied instead of NTOPSIS to assess and compare six supply chain risks: external risks, industry risks, organizational risks, operational risks, supply-side risks, and demand-side risks.

Fahmi et al. [30] defined the concept of triangular neutrosophic cubic linguistic hesitant fuzzy numbers, and proposed related basic operations (including arithmetic averaging and geometric averaging) and the scoring function for evaluating and comparing investment options. As a result, even if the collected data were expressed in much complicated neutrosophic sets, they were converted into crisp values using the scoring function. They also discussed the potential of the proposed methodology for applications in treating various diseases in medical patients.

Irvanizam et al. [31] defined the concept of a bipolar neutrosophic set to model information with vagueness, complexity, and uncertainty. Two aggregation functions for bipolar neutrosophic sets were also proposed. They then proposed a new extended evaluation based on the distance from average solution (EDAS) method based on bipolar neutrosophic sets for selecting small and medium-sized joint business groups and providing them with venture capital assistance, in which the positive distance and the negative distance from each alternative to the average ideal solution were measured to determine the appraisal score of the alternative, based on which the ranking results were obtained.

Reig-Mullor and Salas-Molina [32] defined nonlinear membership functions of neutrosophic sets for assessing supply chain risks in the automotive industry supply in Pakistan, considering three criteria: supply chain resilience, supply chain agility, and supply chain robustness. They also extended α cuts to (α, β, γ) cuts, and considered the mean, variance, and skewness of the possibility function to design a new score function. They applied neutrosophic AHP to derive the priorities of criteria, before which neutrosophic pairwise comparison results were converted into crisp values using the proposed score function. Subsequently, NTOPSIS was used to assess supply chain risks in the automotive industry supply in Pakistan. However, due to the deneutrosophication beforehand, crisp TOPSIS was indeed applied.

Zhong et al. [33] conducted an interval-valued neutrosophic stepwise weight assessment ratio analysis (SWARA) to evaluate the effects of various aspects of knowledge management on the overall performance of a supply chain, in which the relative priorities of criteria were expressed with interval-valued neutrosophic sets. House of quality (HOQ) was also applied to map experts' requirements to criteria or sub-criteria. As in the previous studies, neutrosophic judgments were converted into crisp values before further processing to simplify the required computation.

Irvanizam and Zahara [34] proposed a new ranking of alternatives through functional mapping of criterion sub-intervals into the single interval (RAFSI) method based on single-valued trapezoidal neutrosophic numbers for healthcare service quality evaluation. RAFSI aimed to eliminate the issue of rank reversal. Their methodology consisted of two features. First, the ordinal priority approach (OPA) was devised to determine the weights of criteria. In addition, new harmonic mean and arithmetic mean methods were proposed to yield criterion functions in RAFSI. The function criteria for all alternatives were converted into crisp values using a score function to be compared.

The existing neutrosophic collaborative intelligence methods are compared in Table 1.2.

Figure 1.8 provides statistics on the fields where NCI methods are most frequently applied. Fields in which NCI methods are most frequently applied include neutrosophic collaborative decision-making, neutrosophic collaborative evaluation, neutrosophic collaborative management, and neutrosophic collaborative learning. In contrast, neutrosophic collaborative classification, neutrosophic collaborative recommendation, and neutrosophic collaborative forecasting are fields where NCI methods are rarely applied. Therefore, the application of NCI methods is an urgent task and an opportunity for these fields.

Table 1.2 Comparison of existing neutrosophic collaborative intelligence methods

Reference	Purpose	Type	Participant	Methods	Aggregation method
Chiu and Chen [23]	Cultural difference impact mitigation	Homogeneous	Human	• Neutrosophic AHP • NTOPSIS	NWI
Ihsan et al. [28]	Workstation computer evaluation	Homogeneous	Human	• Neutrosophic weighted average	–
Basset et al. [25]	Strategies selection	Homogeneous	Human	• Neutrosophic AHP • SWOT	Arithmetic averaging
Junaid et al. [29]	Supply chain risk assessment	Homogeneous	Human	• Neutrosophic AHP • TOPSIS	Arithmetic averaging
Fahmi et al. [30]	Investment option comparison	Homogeneous	Human	• Arithmetic averaging • Geometric averaging	–
Irvanizam et al. [31]	Small and medium-sized joint business group selection	Homogeneous	Human	• Neutrosophic EDAS	–
Reig-Mullor and Salas-Molina [32]	Supply chain risks assessment	Homogeneous	Human	• Neutrosophic AHP • NTOPSIS	–
Zhong et al. [33]	Knowledge management effect evaluation	Homogeneous	Human	• Interval-valued neutrosophic SWARA • HOQ	–
Irvanizam and Zahara [34]	Healthcare service quality evaluation	Homogeneous	Human	• Neutrosophic RAFSI	OPA

1.5.1 Organization of This Book

NCI is an extension of FCI. A significant difference between NCI and FCI is the indeterminacy held by the participants, which leaves room for further discussion on the applicability of the aggregation result. For example, in multi-criteria decision-making, a NCI approach can handle decision makers' indeterminacy by incorporating the concept of neutrosophic sets, which is particularly useful when these decision makers do not need to rush to take the corresponding actions. Like FCI, NCI strives to retain most of the uncertainty in the reasoning process until the final step, where it

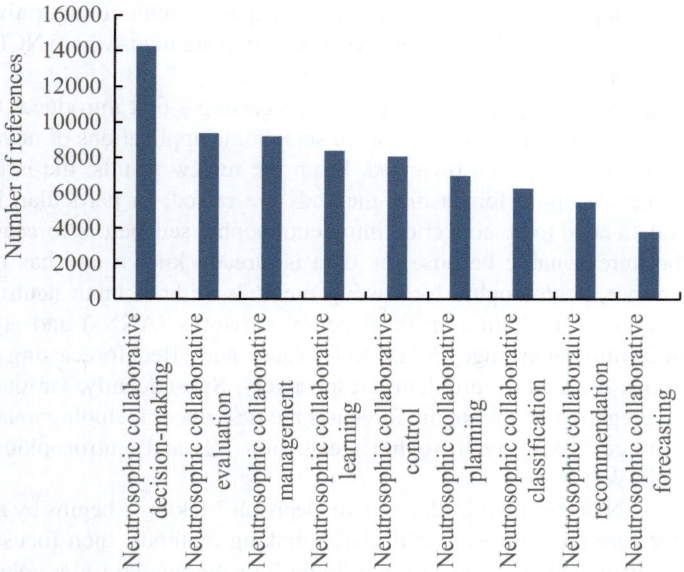

Fig. 1.8 Number of references about NCI applications in various fields until 2026 (data source: Google Scholar)

is deneutrosophicated, to avoid missing the actual value (i.e., the overall consensus), which is distinct from existing neutrosophic set applications and has not been fully addressed. This book is intended to provide technical details about neutrosophic collaborative intelligence with an introduction to the methodology, system architecture, and applications. These details should be of great interest to researchers as well as practicing managers and engineers in industrial engineering, mechanical engineering, medical engineering, operations management, information engineering and management, and others.

In specific, the outline of the present book is structured as follows.

Because NCI evolved from FCI, this chapter first introduces FCI and FCSs, and shows that FCSs are distinct from existing fuzzy ensembles, fuzzy inference systems, fuzzy weighted averaging, fuzzy geometric mean, etc. NCI systems are also expected to differ from existing neutrosophic ensembles, neutrosophic inference systems, and neutrosophic weighted averaging. Subsequently, neutrosophic sets and operations are defined, followed by the formal introduction of NCI and NCSs. Existing neutrosophic collaborative intelligence and systems are then categorized along four dimensions. An implementation process for NCI and NCS is also provided. The chapter then provides examples of various types of neutrosophic collaborative intelligence and systems. Fields in which NCI methods are most frequently applied include neutrosophic collaborative decision-making, neutrosophic collaborative evaluation, neutrosophic collaborative management, and neutrosophic collaborative learning. In contrast, neutrosophic collaborative classification, neutrosophic collaborative recommendation, and neutrosophic collaborative forecasting are fields where NCI methods are rarely applied.

Chapter 2—Neutrosophic Collaborative Forecasting—first introduces the arithmetic and logic operations of neutrosophic sets. Some applications of neutrosophic forecasting methods are then reviewed. From the review results, the issues faced by existing neutrosophic forecasting methods are raised. In particular, historical time-series data need to be converted into neutrosophic sets before forecasting, but this may be unreasonable because the data is already known and has no uncertainty. Therefore, neutrosophic forecasting methods without input neutrosophication are introduced, including artificial neural networks (ANNs) and autoregressive integrated moving average (ARIMA) for cause-and-effect forecasting problems and time-series forecasting problems, respectively. Subsequently, various mechanisms for aggregating neutrosophic forecasts are described, including neutrosophic weighted average (NWA), neutrosophic intersection (NI), and neutrosophic weighted intersection (NWI).

Chapter 3—Neutrosophic Collaborative Decision-Making—begins by reviewing existing literature on neutrosophic decision-making methods, then focuses on the universality of embedding neutrosophic logic into the analytic hierarchy process (AHP) —the technique for order preference by similarity to an ideal solution (TOPSIS). Next, the chapter describes the process of neutrosophic collaborative decision-making. This process begins with decision makers expressing their neutrosophic judgments on the relative weights of various criteria. Based on the neutrosophic judgment matrix, various methods can be applied to derive the criterion weights, such as t neutrosophic geometric mean, neutrosophic extent analysis, deneutrosophication + arithmetic mean, and deneutrosophication + eigen analysis. Finally, this chapter introduces the neutrosophic TOPSIS and its applications. The decision results of multiple decision makers can be aggregated using techniques such as neutrosophic weighted average and neutrosophic weighted intersection. However, aggregating the neutrosophic judgments of multiple decision makers, the indeterminacy of the aggregation result is usually the highest. To resolve this problem, a neutrosophic judgment matrix can be decomposed into several neutrosophic sub-judgment matrixes. Each neutrosophic sub-judgment matrix has a lower indeterminacy than the original neutrosophic judgment matrix.

Chapter 4—Neutrosophic Collaborative Classification—begins by introducing the concept of neutrosophic classification and reviewing existing neutrosophic classification methods. Most existing neutrosophic classification methods convert input attributes into neutrosophic sets before classification. However, this treatment can be problematic because these input attributes may not possess uncertainty. To address this shortcoming, this chapter introduces three neutrosophic classification methods that use the crisp values of example attributes as inputs: neutrosophic c-means clustering, neutrosophic classification trees, and neutrosophic deep neural networks for classification. After describing the implementation process of such neutrosophic classification methods, this chapter introduce the three neutrosophic classification methods one by one. Subsequently, this chapter proposes a neutrosophic collaborative classification method that aggregates the classification results of multiple neutrosophic classification methods.

References

1. T.C.T. Chen, H. Katsuhiro, *Explainable Fuzzy Collaborative Intelligence: Methodology, System Architecture, and Applications in Engineering* (Springer International Publishing, 2026)
2. X. Zhu, D. Wang, W. Pedrycz, Z. Li, Horizontal federated learning of Takagi-Sugeno fuzzy rule-based models. IEEE Trans. Fuzzy Syst. **30**(9), 3537–3547 (2021)
3. X. Zhang, C. Bo, F. Smarandache, J. Dai, New inclusion relation of neutrosophic sets with applications and related lattice structure. Int. J. Mach. Learn. Cybern. **9**(10), 1753–1763 (2018)
4. W. Pedrycz, Collaborative fuzzy clustering. Pattern Recognit. Lett. **23**, 1675–1686 (2002)
5. W. Pedrycz, Collaborative architectures of fuzzy modeling. Lect. Notes Comput. Sci. **5050**, 117–139 (2008)
6. T.C.T. Chen, K. Honda, *Fuzzy Collaborative Forecasting and Clustering: Methodology, System Architecture, and Applications* (Springer International Publishing, 2020)
7. H. Yuan, P. Zhao, J. Fan, J. Fang, G. Liu, V.S. Sheng, Fuzzy Collaborative Reasoning. Proc. AAAI Conf. Artif. Intell. **39**(12), 13124–13132 (2025)
8. T.C.T. Chen, Y.C. Wang, Evaluating innovative future robotic applications in manufacturing using a fuzzy collaborative intelligence approach. Int. J. Adv. Manuf. Technol. **130**(11), 6027–6041 (2024)
9. T. Chen, The FLR–PCFI–RBF approach for accurate and precise WIP level forecasting. Int. J. Adv. Manuf. Technol. **63**(9), 1217–1226 (2012)
10. T.C.T. Chen, Y.C. Wang, C.W. Lin, A fuzzy collaborative forecasting approach considering experts' unequal levels of authority. Appl. Soft Comput. **94**, 106455 (2020)
11. J. Zhang, L. Jin, Y. Wang, Collaborative control for multimanipulator systems with fuzzy neural networks. IEEE Trans. Fuzzy Syst. **31**(4), 1305–1314 (2022)
12. A. Sathyan, O. Ma, Collaborative control of multiple robots using genetic fuzzy systems. Robotica **37**(11), 1922–1936 (2019)
13. Y. Jiang, Z. Deng, F.L. Chung, S. Wang, Realizing two-view TSK fuzzy classification system by using collaborative learning. IEEE Trans. Syst., Man, Cybern.: Syst. **47**(1), 145–160 (2016)
14. I. Matazi, A. Bennane, R. Messoussi, R. Touahni, I. Oumaira, R. Korchiyne, Multi-agent system based on fuzzy logic for e-learning collaborative system. in *2018 International Symposium on Advanced Electrical and Communication Technologies* (2018), pp. 1–7
15. O. Ayadi, N. Halouani, F. Masmoudi, A fuzzy collaborative assessment methodology for partner trust evaluation. Int. J. Intell. Syst. **31**(5), 488–501 (2016)
16. Y. Bao, W. Lei, W. Zhang, Y. Zhan, QoE collaborative evaluation method based on fuzzy clustering heuristic algorithm. Springerplus **5**(1), 1008 (2016)
17. S. Alias, D. Mohamad, A review on neutrosophic set and its development. Discov. Math. (Menemui Matematik) **39**(2), 61–69 (2017)
18. M. Abdel-Basset, A. Gamal, G. Manogaran, L.H. Son, H.V. Long, A novel group decision making model based on neutrosophic sets for heart disease diagnosis. Multimed. Tools Appl. **79**, 9977–10002 (2020)
19. T.S. Haque, A. Chakraborty, S. Alam, E-learning app selection multi-criteria group decision-making problem using Einstein operator in linguistic trapezoidal neutrosophic environment. Knowl. Inf. Syst. **67**, 2481–2519 (2025)
20. M. Saeed, F. Razaq, I. Tariq, I.S.U. Din, Navigating complexity: a multidimensional neutrosophic fuzzy hypersoft approach to empowering decision-makers. J. Ind Intell **1**(4), 219–228 (2023)

21. J. Zhan, M. Akram, M. Sitara, Novel decision-making method based on bipolar neutrosophic information. Soft. Comput. **23**, 9955–9977 (2019)
22. M.A. Dasan, E. Bementa, M. Aslam, V.L. Flower, Multi-attribute decision-making problem in career determination using single-valued neutrosophic distance measure. Complex Intellig. Syst. **10**(4), 5411–5425 (2024)
23. M.C. Chiu, T.C.T. Chen, Prioritizing measures to mitigate the impact of cultural differences on semiconductor supply chain localization: a neutrosophic collaborative intelligence approach. Cogn. Comput. **17**(4), 125 (2025)
24. E. Bolturk, C. Kahraman, A novel interval-valued neutrosophic AHP with cosine similarity measure. Soft. Comput. **22**, 4941–4958 (2018)
25. M.A. Basset, M. Mohamed, A.K. Sangaiah, V. Jain, An integrated neutrosophic AHP and SWOT method for strategic planning methodology selection. Benchmarking: Int. J. **25**(7), 2546–2564 (2018)
26. M. Abdel-Basset, V. Chang, M. Mohamed, F. Smarandache, A refined approach for forecasting based on neutrosophic time series. Symmetry **11**(4), 457 (2019)
27. M. Abdel-Basset, M. Mohamed, Y. Zhou, I. Hezam, Multi-criteria group decision making based on neutrosophic analytic hierarchy process. J. Intell. Fuzzy Syst. **33**(6), 4055–4066 (2017)
28. M. Ihsan, M. Saeed, A.U. Rahman, Optimizing hard disk selection via a fuzzy parameterized single-valued neutrosophic soft set approach. J. Oper. Strat. Anal. **1**(2), 62–69 (2023)
29. M. Junaid, Y. Xue, M.W. Syed, J.Z. Li, M. Ziaullah, A neutrosophic ahp and topsis framework for supply chain risk assessment in automotive industry of Pakistan. Sustainability **12**(1), 154 (2019)
30. A. Fahmi, M. Aslam, M. Riaz, New approach of triangular neutrosophic cubic linguistic hesitant fuzzy aggregation operators. Granul. Comput. **5**(4), 527–543 (2020)
31. I. Irvanizam, I. Syahrini, N.N. Zi, N. Azzahra, M. Iqbal, M. Marzuki, M. Subianto, An improved EDAS method based on bipolar neutrosophic set and its application in group decision-making. Appl. Comput. Intell. Soft Comput. **2021**(1), 1474629 (2021)
32. J. Reig-Mullor, F. Salas-Molina, Non-linear neutrosophic numbers and its application to multiple criteria performance assessment. Int. J. Fuzzy Syst. **24**(6), 2889–2904 (2022)
33. J. Zhong, H. Cheng, H. Gholami, L.T. Letchumanan, Ş Toptancı, Supply chain performance: a novel integrated decision-making model. Manag. Decis. **61**(10), 3053–3081 (2023)
34. I. Irvanizam, N. Zahara, An improved RAFSI method based on single-valued trapezoidal neutrosophic number and its harmonic and arithmetic mean operators for healthcare service quality evaluation. Expert Syst. Appl. **248**, 123343 (2024)

Chapter 2
Neutrosophic Collaborative Forecasting

Abstract This chapter first introduces the arithmetic and logic operations of neutrosophic sets. Some applications of neutrosophic forecasting methods are then reviewed. From the review results, the issues faced by existing neutrosophic forecasting methods are raised. In particular, historical time-series data need to be converted into neutrosophic sets before forecasting, but this may be unreasonable because the data is already known and has no uncertainty. Therefore, neutrosophic forecasting methods without input neutrosophication are introduced, including artificial neural networks (ANNs) and autoregressive integrated moving average (ARIMA) for cause-and-effect forecasting problems and time-series forecasting problems, respectively. Subsequently, various mechanisms for aggregating neutrosophic forecasts are described, including neutrosophic weighted average (NWA), neutrosophic intersection (NI), and neutrosophic weighted intersection (NWI).

Keywords Forecasting · Triangular neutrosophic number · Neutrosophic weighted average · Neutrosophic intersection · Neutrosophic weighted intersection · Neutrosophic collaborative forecasting

2.1 Neutrosophic Sets and Operations

Most commonly used neutrosophic sets in forecasting include triangular neutrosophic sets [1], single-valued neutrosophic sets (in contrast to interval-valued neutrosophic sets) [2], trapezoidal neutrosophic sets [3], etc.

Definition 2.1 [1] A triangular neutrosophic number (TNN) $\tilde{A} = (A_1, A_2, A_3); \alpha_{\tilde{A}}, \theta_{\tilde{A}}, \beta_{\tilde{A}}$ is a special neutrosophic set in **R** with the following truth, indeterminacy, and falsity membership functions:

T.-C.T. Chen, *Neutrosophic Collaborative Intelligence and Systems*,
SpringerBriefs in Applied Sciences and Technology,
https://doi.org/10.1007/978-3-032-34267-6_2

$$T_{\tilde{A}}(x) = \begin{cases} \alpha_{\tilde{A}}\left(\frac{x-A_1}{A_2-A_1}\right) & if \quad A_1 \le x < A_2 \\ \alpha_{\tilde{A}}\left(\frac{A_3-x}{A_3-A_2}\right) & if \quad A_2 \le x < A_3 \\ 0 & otherwise \end{cases} \tag{2.1}$$

$$I_{\tilde{A}}(x) = \begin{cases} \frac{A_2-x+\theta_{\tilde{A}}(x-A_1)}{A_2-A_1} & if \quad A_1 \le x < A_2 \\ \frac{x-A_2+\theta_{\tilde{A}}(A_3-x)}{A_3-A_2} & if \quad A_2 \le x < A_3 \\ 0 & otherwise \end{cases} \tag{2.2}$$

$$F_{\tilde{A}}(x) = \begin{cases} \frac{A_2-x+\beta_{\tilde{A}}(x-A_1)}{A_2-A_1} & if \quad A_1 \le x < A_2 \\ \frac{x-A_2+\beta_{\tilde{A}}(A_3-x)}{A_3-A_2} & if \quad A_2 \le x < A_3 \\ 0 & otherwise \end{cases} \tag{2.3}$$

where $\alpha_{\tilde{A}}, \theta_{\tilde{A}}$, and $\beta_{\tilde{A}}$ indicate the highest truth membership, the lowest indeterminacy membership, and the lowest falsity membership, respectively.

Some arithmetic operations on TNNs are described as follows, using product t-norm (or Lukasiewicz t-norm) and t-conorm (Lukasiewicz t-conorm) [4], which can effectively avoid the drawback of standard t-norm and t-conorm based on intersection (minimum) and union (maximum) operators, that is, the judgment with the highest (or lowest) membership dominates the derivation results.

Theorem 2.1 (TNN Arithmetic Operations) [4]

- Addition: $\tilde{A}(+)\tilde{B} = (A_1 + B_1, A_2 + B_2, A_3 + B_3); \alpha_{\tilde{A}} + \alpha_{\tilde{B}} - \alpha_{\tilde{A}}\alpha_{\tilde{B}}, \theta_{\tilde{A}}\theta_{\tilde{B}}, \beta_{\tilde{A}}\beta_{\tilde{B}}$
- Subtraction: $\tilde{A}(-)\tilde{B} = (A_1 - B_3, A_2 - B_2, A_3 - B_1); \alpha_{\tilde{A}} + \alpha_{\tilde{B}} - \alpha_{\tilde{A}}\alpha_{\tilde{B}}, \theta_{\tilde{A}}\theta_{\tilde{B}}, \beta_{\tilde{A}}\beta_{\tilde{B}}$
- Inverse: $\frac{1}{\tilde{A}} \cong \left(\frac{1}{A_3}, \frac{1}{A_2}, \frac{1}{A_1}\right); \alpha_{\tilde{A}}, \theta_{\tilde{A}}, \beta_{\tilde{A}}$ if $A_1 \ge 0$
- Multiplication: $\tilde{A}(\times)\tilde{B} \cong (A_1B_1, A_2B_2, A_3B_3); \alpha_{\tilde{A}}\alpha_{\tilde{B}}, \theta_{\tilde{A}} + \theta_{\tilde{B}} - \theta_{\tilde{A}}\theta_{\tilde{B}}, \beta_{\tilde{A}} + \beta_{\tilde{B}} - \beta_{\tilde{A}}\beta_{\tilde{B}}$ if $A_1, B_1 \ge 0$
- Division: $\tilde{A}(/)\tilde{B} \cong (A_1/B_3, A_2/B_2, A_3/B_1); \alpha_{\tilde{A}}\alpha_{\tilde{B}}, \theta_{\tilde{A}} + \theta_{\tilde{B}} - \theta_{\tilde{A}}\theta_{\tilde{B}}, \beta_{\tilde{A}} + \beta_{\tilde{B}} - \beta_{\tilde{A}}\beta_{\tilde{B}}$ if $B_1 \ge 0$
- Square root: $\sqrt{\tilde{A}} \cong \left(\sqrt{A_1}, \sqrt{A_2}, \sqrt{A_3}\right); \alpha_{\tilde{A}}, \theta_{\tilde{A}}, \beta_{\tilde{A}}$ if $A_1 \ge 0$

Some aggregation operators for TNNs are introduced below.

Theorem 2.2 (WA Based on TNNs).

$$\begin{aligned} w_1\tilde{A}(+)w_2\tilde{B} &= (w_1A_1 + w_2B_1, w_1A_2 + w_2B_2, w_1A_3 + w_2B_3); \\ &\quad \alpha_{\tilde{A}} + \alpha_{\tilde{B}} - \alpha_{\tilde{A}}\alpha_{\tilde{B}}, \theta_{\tilde{A}}\theta_{\tilde{B}}, \beta_{\tilde{A}}\beta_{\tilde{B}} \end{aligned} \tag{2.4}$$

Neutrosophic weighted average (NWA) where both the weights and values of variables are TNNs are described as follows.

Theorem 2.3 (NWA Based on TNNs).

$$\begin{aligned} &\tilde{w}_1(\times)\tilde{A}(+)\tilde{w}_2(\times)\tilde{B} = (w_{11}A_1 + w_{21}B_1, w_{12}A_2 + w_{22}B_2, w_{13}A_3 + w_{23}B_3); \\ &\alpha_{\tilde{w}_1}\alpha_{\tilde{A}} + \alpha_{\tilde{w}_2}\alpha_{\tilde{B}} - \alpha_{\tilde{w}_1}\alpha_{\tilde{A}}\alpha_{\tilde{w}_2}\alpha_{\tilde{B}}, \left(\theta_{\tilde{w}_1} + \theta_{\tilde{A}} - \theta_{\tilde{w}_1}\theta_{\tilde{A}}\right)\left(\theta_{\tilde{w}_2} + \theta_{\tilde{B}} - \theta_{\tilde{w}_2}\theta_{\tilde{B}}\right), \\ &\left(\beta_{\tilde{w}_1} + \beta_{\tilde{A}} - \beta_{\tilde{w}_1}\beta_{\tilde{A}}\right)\left(\beta_{\tilde{w}_2} + \beta_{\tilde{B}} - \beta_{\tilde{w}_2}\beta_{\tilde{B}}\right) \end{aligned} \tag{2.5}$$

If the weights are fuzzy numbers rather than neutrosophic numbers,

$$\begin{aligned} w_1\tilde{A}(+)w_2\tilde{B} &= (w_{11}A_1 + w_{21}B_1, w_{12}A_2 + w_{22}B_2, w_{13}A_3 + w_{23}B_3); \\ &\alpha_{\tilde{A}} + \alpha_{\tilde{B}} - \alpha_{\tilde{A}}\alpha_{\tilde{B}}, \theta_{\tilde{A}}\theta_{\tilde{B}}, \beta_{\tilde{A}}\beta_{\tilde{B}} \end{aligned} \tag{2.6}$$

In addition, the neutrosophic intersection (NI) of two TNNs are derived as follows.

Theorem 2.4 (NI Based on TNNs).

$$\widetilde{NI}\left(\tilde{A}, \tilde{B}\right) = [\max(A_1, B_1), \min(A_3, B_3)]; \alpha_{\tilde{A}} + \alpha_{\tilde{B}} - \alpha_{\tilde{A}}\alpha_{\tilde{B}}, \theta_{\tilde{A}}\theta_{\tilde{B}}, \beta_{\tilde{A}}\beta_{\tilde{B}} \tag{2.7}$$

The neutrosophic union (NU) of two TNNs is derived as follows.

Theorem 2.5 (NU Based on TNNs).

$$\begin{aligned} \widetilde{NU}\left(\tilde{A}, \tilde{B}\right) &= [\min(A_1, B_1), \max(A_3, B_3)]; \\ &\alpha_{\tilde{A}}\alpha_{\tilde{B}}, \theta_{\tilde{A}} + \theta_{\tilde{B}} - \theta_{\tilde{A}}\theta_{\tilde{B}}, \beta_{\tilde{A}} + \beta_{\tilde{B}} - \beta_{\tilde{A}}\beta_{\tilde{B}} \end{aligned} \tag{2.8}$$

NI and NU are applicable if there are multiple forecasters, or a forecaster makes multiple forecasts from various viewpoints, then these forecasts need to be aggregated [5, 6].

Another way to calculate the arithmetic and logic operations of TNNs is standard t-norm (or Gödel–Dummett t-norm) and standard t-conorm (or Gödel–Dummett t-conorm), described as follows.

Theorem 2.6 (TNN Arithmetic Operations) [7, 8]

- Addition:
$\tilde{A}(-)\tilde{B} =$
$\left\langle (A_1 - B_3,\ A_2 - B_2, A_3 - B_1);\ \min\left(\alpha_{\tilde{A}},\ \alpha_{\tilde{B}}\right), \max\left(\theta_{\tilde{A}},\ \theta_{\tilde{B}}\right), \max(\beta_{\tilde{A}},\ \beta_{\tilde{B}})\right\rangle$
- Subtraction:
$\tilde{A}(-)\tilde{B} =$
$\left\langle (A_1 - B_3,\ A_2 - B_2, A_3 - B_1);\ \min\left(\alpha_{\tilde{A}},\ \alpha_{\tilde{B}}\right), \max\left(\theta_{\tilde{A}},\ \theta_{\tilde{B}}\right), \max(\beta_{\tilde{A}},\ \beta_{\tilde{B}})\right\rangle$
- Inverse: $\frac{1}{\tilde{A}} \cong \left(\frac{1}{A_3}, \frac{1}{A_2}, \frac{1}{A_1}\right); \alpha_{\tilde{A}}, \theta_{\tilde{A}}, \beta_{\tilde{A}}$ if $A_1 \geq 0$
- Multiplication:
$\tilde{A}(\times)\tilde{B} \cong \left\langle (A_1B_1,\ A_2B_2, A_3B_3); \min\left(\alpha_{\tilde{A}},\ \alpha_{\tilde{B}}\right), \max\left(\theta_{\tilde{A}},\ \theta_{\tilde{B}}\right), \max(\beta_{\tilde{A}},\ \beta_{\tilde{B}})\right\rangle$ if $A_1, B_1 \geq 0$
- Division:
$\tilde{A}(/)\tilde{B} \cong \left\langle (A_1/B_3,\ A_2/B_2, A_3/B_1); \min\left(\alpha_{\tilde{A}},\ \alpha_{\tilde{B}}\right), \max\left(\theta_{\tilde{A}},\ \theta_{\tilde{B}}\right), \max(\beta_{\tilde{A}},\ \beta_{\tilde{B}})\ \right\rangle$ if $B_1 \geq 0$
- Square root: $\sqrt{\tilde{A}} \cong \left(\sqrt{A_1}, \sqrt{A_2}, \sqrt{A_3}\right); \alpha_{\tilde{A}}, \theta_{\tilde{A}}, \beta_{\tilde{A}}$ if $A_1 \geq 0$

Some aggregation operators for TNNs based on standard t-norm and standard t-conorm are introduced below.

Theorem 2.7 (WA Based on TNNs).

$$\begin{aligned} &w_1\tilde{A}(+)w_2\tilde{B} = (w_1A_1 + w_2B_1, w_1A_2 + w_2B_2, w_1A_3 + w_2B_3);\\ &\min\left(\alpha_{\tilde{A}}, \alpha_{\tilde{B}}\right), \max\left(\theta_{\tilde{A}}, \theta_{\tilde{B}}\right), \max(\beta_{\tilde{A}}, \beta_{\tilde{B}}) \end{aligned} \tag{2.9}$$

Theorem 2.8 (NWA Based on TNNs).

$$\begin{aligned} &\tilde{w}_1(\times)\tilde{A}(+)\tilde{w}_2(\times)\tilde{B} = \begin{pmatrix} w_{11}A_1 + w_{21}B_1, w_{12}A_2 + w_{22} \\ B_2, w_{13}A_3 + w_{23}B_3 \end{pmatrix};\\ &\min\left(\alpha_{\tilde{w}_1}, \alpha_{\tilde{w}_2}, \alpha_{\tilde{A}}, \alpha_{\tilde{B}}\right), \max\left(\alpha_{\tilde{w}_1}, \alpha_{\tilde{w}_2}, \alpha_{\tilde{A}}, \alpha_{\tilde{B}}\right),\\ &\max\left(\alpha_{\tilde{w}_1}, \alpha_{\tilde{w}_2}, \alpha_{\tilde{A}}, \alpha_{\tilde{B}}\right), \end{aligned} \tag{2.10}$$

Theorem 2.9 (NI Based on TNNs).

$$\begin{aligned} &\widetilde{NI}\left(\tilde{A}, \tilde{B}\right) = [\max(A_1, B_1), \min(A_3, B_3)];\\ &\min\left(\alpha_{\tilde{A}}, \alpha_{\tilde{B}}\right), \max\left(\theta_{\tilde{A}}, \theta_{\tilde{B}}\right), \max(\beta_{\tilde{A}}, \beta_{\tilde{B}}) \end{aligned} \tag{2.11}$$

Theorem 2.10 (NU Based on TNNs).

$$\begin{aligned} &\widetilde{NU}\left(\tilde{A}, \tilde{B}\right) = [\min(A_1, B_1), \max(A_3, B_3)];\\ &\max\left(\alpha_{\tilde{A}}, \alpha_{\tilde{B}}\right), \min\left(\theta_{\tilde{A}}, \theta_{\tilde{B}}\right), \min(\beta_{\tilde{A}}, \beta_{\tilde{B}}) \end{aligned} \tag{2.12}$$

2.2 Existing Neutrosophic Forecasting Methods

Abdel-Basset et al. [9] proposed a neutrosophic time-series forecasting method based on the theory of neutrosophic sets to predict student enrollment rates and stock market indices. In their methodology, to convert historical time-series data into neutrosophic values, the range of the target time series was determined. Then, this range was divided into several intervals. For each interval, a neutrosophic set was defined, and its three membership functions were subjectively determined by domain experts. Subsequently, the data for each period was mapped to a neutrosophic set, and neutrosophic logical relations (NLRs) were established to fit the relationship between neutrosophic sets of two consecutive periods. However, whether converting historical time-series data without uncertainty into neutrosophic sets was reasonable remained controversial.

For a similar purpose, Guan et al. [10] adopted a more objective method to convert historical time-series data into neutrosophic sets. First, the upward trend of each period was mapped to the truth membership of the neutrosophic set for that period, while the downward trend was mapped to the falsity membership. The information entropy of higher order fluctuation time series was mapped to the indeterminacy membership of the neutrosophic set. Although this approach seemed more reasonable, a question still remained: why was it necessary to convert historical time-series data without uncertainty into neutrosophic sets with uncertainty?

Guan et al. [11] extended previous work to handle multi-attribute time series that were closely related to each other, such as stock market volume, stock market index, and daily volatility. They used neutrosophic soft sets (NSSs) to represent these closely related attributes. Similar to Guan et al. [10], they considered the upward trend, downward trend, and volatility of each period to determine the three memberships of the NSS for that period. Same as Abdel-Basset et al. [9], they also established NLRs to predict the values of these attributes for future periods.

Chen and Lin [12] proposed a neutrosophic reasoning approach for trustworthy predictive maintenance in smart manufacturing. In the neutrosophic reasoning approach, a deep neutrosophic neural network (DNNN) was constructed and trained to predict the next time to failure of a machine, so as to arrange preventive maintenance in advance. By considering the overall prediction accuracy and the local prediction accuracy and learning sufficiency of each case, the three memberships of the neutrosophic time-to-failure forecast were derived. If the indeterminacy of the neutrosophic time-to-failure forecast exceeded a threshold, the prediction result was considered not trustworthy, because similar cases were either mispredicted or rarely predicted before. The neutrosophic reasoning approach was applied to a case of a computer numeric control (CNC) tool.

The existing neutrosophic forecasting methods are compared in Table 2.1.

Table 2.1 Comparison of existing neutrosophic forecasting methods

Reference	Forecasting target	Forecasting method	Type of neutrosophic set
Abdel-Basset et al. [9]	Student enrollment rates, Stock market indices	NLR	TNN
Guan et al. [10]	Time series	Trend analysis, information entropy	NSS
Guan et al. [11]	Stock market volume, stock market index, daily volatility	Trend and volatility analysis	NSS
Chen and Lin [12]	Time to failure	DNNN	TNN

2.3 Problems of Existing Neutrosophic Forecasting Methods

Existing methods in this field are subject to the following issues (see Fig. 2.1):

- Historical time-series data need to be converted into neutrosophic sets, but this may be unreasonable because the data is already known and has no uncertainty.
- The memberships of time-series data are independent of the chosen forecasting method. If the forecasting method is perfect, values for all periods can be exactly predicted, and then there will be no uncertainty. Therefore, the source of uncertainty seems to be the inaccuracy of the time-series forecasting method rather than the data itself, which may be considered in deriving the memberships.
- If the time-series data in two periods are the same, the memberships for them still differ. In other words, they will be represented with different neutrosophic sets.
- It is hard to guarantee that each neutrosophic forecast contains the corresponding actual value, which is a prerequisite for fuzzy or neutrosophic collaborative forecasting methods.
- Most existing methods are for time-series forecasting problems, seldom for cause-and-effect forecasting problems.
- In the next section, a neutrosophic forecasting method without transforming input variables to neutrosophic sets is introduced.

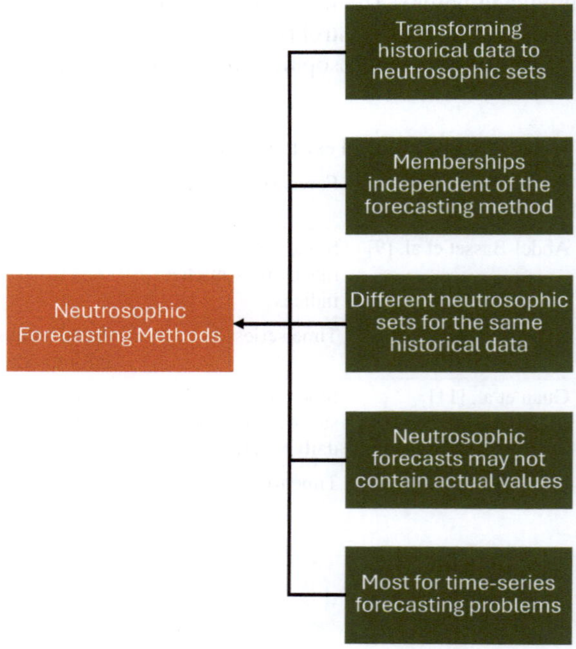

Fig. 2.1 Issues faced by existing neutrosophic forecasting methods

2.4 Neutrosophic Forecasting Methods Without Input Neutrosophication

2.4.1 *Cause-and-Effect Forecasting Problems*

The memberships of a neutrosophic forecast can be derived by considering forecasting inaccuracy and learning insufficiency. Take a job cycle time forecasting problem as an example, which is a cause-and-effect/casual forecasting problem. An artificial neural network (ANN) is constructed to predict the cycle time of a job from its K attributes [13]:

$$\hat{y}_j = ANN(\{x_{jk} | k = 1 \sim K\}) \tag{2.13}$$

where $\hat{y}_j$ is the cycle time forecast of job j; x_{jk} is the k-th attribute of this job. The operations in the ANN are described below

$$\hat{y}_j = \sum_{l=1}^{L} w_l^o h_{jl} - \theta^o \tag{2.14}$$

where w_l^o is the connection weight between the l-th hidden-layer node and the output node. h_{jl} is the output from the l-th hidden-layer node for job j. θ^o is the threshold on the output node. In Eq. (2.14), the activation function is the linear function (purelin). h_{jl} is derived as

$$h_{jl} = \frac{2}{1 + e^{-2\left(\sum_{k=1}^{K} w_{kl}^{(h)} x_{jk} - \theta_l^{(h)}\right)}} - 1 \tag{2.15}$$

where $w_{kl}^{(h)}$ is the connection weight between the k-th input node and the l-th hidden-layer node. $\theta_l^{(h)}$ is the threshold on the l-th hidden-layer node. Here the activation function is the hyperbolic tangent function (tanh). The forecasting accuracy is measured in terms of root mean squared error (RMSE):

$$RMSE = \sqrt{\frac{\sum_{j=1}^{n} (y_j - \hat{y}_j)^2}{n}} \tag{2.16}$$

where y_j is the actual value.

The truth membership of $\hat{y}_j$ indicates the possibility that y_j is equal to $\hat{y}_j$, rather than another value. Therefore, it can be measured by the accuracy of $\hat{y}_j$. However, the training of ANN is to maximize overall forecasting accuracy, while the accuracy for an individual forecast is not guaranteed. Therefore, overall forecasting accuracy, in terms of RMSE, is considered in evaluating the truth membership of each forecast:

- The truth membership of $\hat{y}_j$ increases as RMSE decreases.
- The falsity membership is the opposite.
- The falsity membership of $\hat{y}_j$ increases as RMSE increases.

There are various functions to achieve the desired effects:
(Exponential)

$$\alpha_{\hat{y}_j} = e^{-\rho_1 \cdot RMSE} \tag{2.17}$$

$$\beta_{\hat{y}_j} = 1 - e^{-\rho_1 \cdot RMSE} \tag{2.18}$$

(Logarithmic)

$$\alpha_{\hat{y}_j} = \log\left(\frac{1}{\rho_2 \cdot RMSE + c} + 1\right) \tag{2.19}$$

$$\beta_{\hat{y}_j} = 1 - \log\left(\frac{1}{\rho_2 \cdot RMSE + c} + 1\right) \tag{2.20}$$

where $\rho_1, \rho_2 \in \mathbf{R}^+$ are constants; $c = \frac{1}{e-1}$. The requirements for these functions are

- $\alpha_{\hat{y}_j} = 0$ if RMSE $\rightarrow \infty$
- $\alpha_{\hat{y}_j} = 1$ if RMSE $= 0$.

The RMSE in fitting the training data is used to derive the memberships of forecasts in test data. In addition, the memberships of all forecasts are the same.

Example 2.1 Cycle time-related data have been collected for 200 jobs and are presented in Table 2.2, in which the cycle time of job j, y_j, is to be predicted based on the values of six job attributes, $x_{j1} \sim x_{j6}$,, using an ANN. The first 150 jobs are used as the training data, while the remaining jobs are left for testing. The hidden layer in the ANN has 12 nodes. The Levenberg–Marquardt (LM) algorithm is applied to train the ANN in 20,000 epochs. The target for network convergence is RMSE ≤ 10 (hrs). The MATLAB code for implementing the ANN is shown in Fig. 2.2. After applying the trained ANN to test data, the forecasting results are summarized in Fig. 2.3. The RMSEs for the training and test data are 4.3 and 24.4 (hrs), respectively. By setting $\rho_1 = 0.0105$, so that $\alpha_{\hat{y}_j} = 0.9$ when RMSE $= 10$ (hrs), the truth and falsity memberships of each cycle time forecast in test data are derived as.

$$\alpha_{\hat{y}_j} = e^{-0.0105 \cdot 4.3} = 0.956 \, \forall j = 151 \sim 200.$$

$$\beta_{\hat{y}_j} = 1 - 0.956 = 0.044 \, \forall j = 151 \sim 200.$$

Table 2.2 Cycle time-related data of 200 jobs

j	x_{j1}	x_{j2}	x_{j3}	x_{j4}	x_{j5}	x_{j6} (%)	y_j (hrs)
1	22	1320	165	748	421	85	1298
2	23	1261	169	734	198	81	1098
3	22	1187	159	635	287	93	1122
200	24	1309	175	734	345	81	1276

```
cd F:\NCI
cycletimeforecastingexample=readtable("cycle time forecasting example.xlsx");

cycletimeforecastingexample=table2array(cycletimeforecastingexample);

train_x=cycletimeforecastingexample(1:150,2:7);
train_y=cycletimeforecastingexample(1:150,8:8);
test_x=cycletimeforecastingexample(151:200,2:7);
test_y=cycletimeforecastingexample(151:200,8:8);
train_x=transpose(train_x);
train_y=transpose(train_y);
test_x=transpose(test_x);
test_y=transpose(test_y);

net1=feedforwardnet([12]);
net1.dividefcn='dividetrain';
net1.trainParam.lr=0.1;
net1.trainParam.epochs=20000;
net1.trainParam.goal=10;
net1=train(net1,train_x,train_y);
train_o=net1(train_x);
test_o=net1(test_x);

train_rmse=mean((train_y-train_o).^2)^0.5
test_rmse=mean((test_y-test_o).^2)^0.5

plot(cycletimeforecastingexample(151:200,1:1),test_y,'o',cycletimeforecastingexample(151:200,1:1)
,test_o);
title('Cycle Time forecasting Results');
xlabel('Job #');
ylabel('Cycle time (hrs)');
```

Fig. 2.2 MATLAB code for implementing the ANN

The effects of the value of ρ_1 on the two memberships are illustrated in Figs. 2.4 and 2.5, respectively. A forecaster can choose the value of ρ_1 by requesting the truth membership higher than a pre-specified level or the falsity membership lower than another pre-specified level.

The indeterminacy membership indicates that it is unclear whether $\hat{y}_j$ is a suitable forecast of y_j. In other words, the trustworthiness of $y_j = \hat{y}_j$ is not necessarily high and may be low, due to the following reasons:

- Examples similar to $\mathbf{x_j}$ have rarely been learned.
- The target variables for examples similar to $\mathbf{x_j}$ have not been accurately predicted.

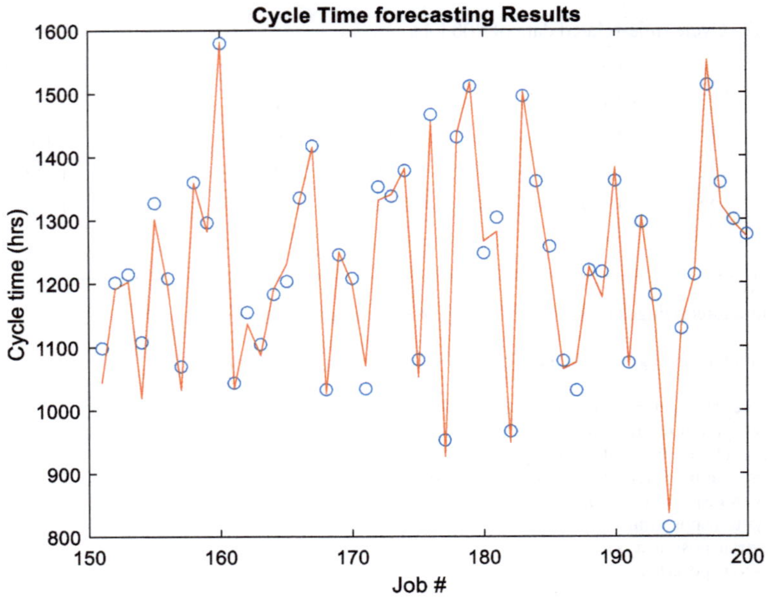

Fig. 2.3 Cycle time forecasting results

Fig. 2.4 Effects of the value of ρ_1 on the truth membership

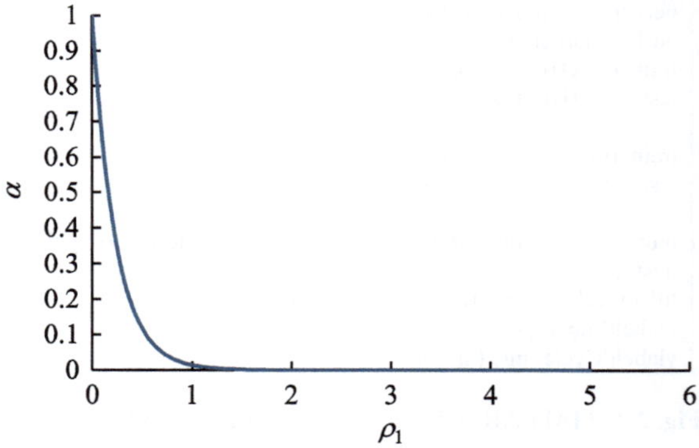

Therefore, the following formula is used to derive the indeterminacy of $\hat{y}_j$:

$$\theta_{\hat{y}_j} = 1 - e^{-\rho_3 \cdot \frac{RMSE(prox(j))}{|prox(j)|}} \tag{2.21}$$

where $\rho_3 \in \mathbf{R}^+$ is a constant; *prox*(*j*) measures the neighborhood crowding degree of job *j* in test data, reflecting the learning sufficiency for job *j*. $l \in prox(j)$ if $dist(l, j) \leq \in$, where job *l* belongs to the training data, as illustrated in Fig. 2.6. The distance between two jobs/examples is measured using their Euclidean distance as

Fig. 2.5 Effects of the value of ρ_1 on the falsity membership

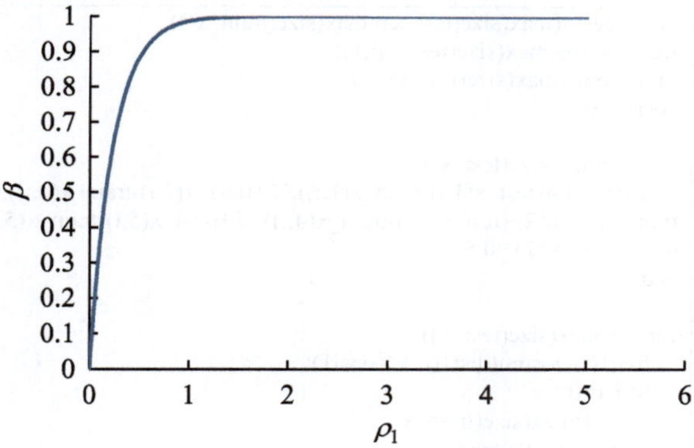

Fig. 2.6 Proximity of an example

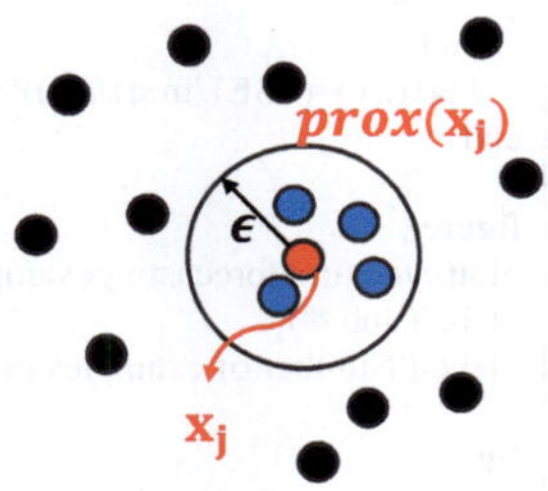

$$dist(l, j) = \sqrt{\sum_k (x_{lk} - x_{jk})^2} \tag{2.22}$$

$RMSE(prox(j))$ is the local prediction accuracy considering only jobs in $prox(j)$:

$$RMSE(prox(j)) = \sqrt{\frac{\sum_{l \in prox(j)} (y_l - \hat{y}_l)^2}{|prox(j)|}} \tag{2.23}$$

The evaluation results make the cycle time forecast of job j a single-valued neutrosophic set indicated with

$$\tilde{y}_j = \hat{y}_j; \alpha_{\hat{y}_j}, \theta_{\hat{y}_j}, \beta_{\hat{y}_j} \tag{2.24}$$

Example 2.2 For the previous example, the neighborhood crowding degree of each job in test data is evaluated, for which $\in$ is set to 50. The required MATLAB code is provided in Fig. 2.7. The evaluation results are summarized in Fig. 2.8. The local prediction accuracy around each job is shown in Fig. 2.9. By setting $\rho_3 = 0.15$, so that when $RMSE(prox(j)) = 20$ and $|prox(j)| = 1$ then $\theta_{\tilde{y}_j} = 0.95$, the indeterminacy membership of each job in test data is derived and shown in Table 2.3.

The three memberships of all jobs are summarized in Fig. 2.10.

```
dist1=zeros(max(size(test_x)),max(size(train_x)));
freq1=zeros(max(size(test_x)),1);
lda1=zeros(max(size(test_x)),1);
epsi1=50;

for i=1:max(size(test_x))
   dist1(i,:)=((test_x(1,i)-train_x(1,:)).^2+(test_x(2,i)-train_x(2,:)).^2+(test_x(3,i)-
train_x(3,:)).^2+(test_x(4,i)-train_x(4,:)).^2+(test_x(5,i)-train_x(5,:)).^2+(test_x(6,i)-
train_x(6,:)).^2).^0.5;
end

for i=1:max(size(test_x))
   freq1(i,1)=sum(dist1(i,:)<=epsi1);
   SSE1=0;
   for l=1:max(size(train_x))
      if dist1(i,l)<=epsi1
         SSE1=SSE1+(train_y(1,l)-train_o(1,l))^2;
      end
   end
   lda1(i,1)=(SSE1/freq1(i,1))^0.5;
end

figure;
plot(cycletimeforecastingexample(151:200,1:1),freq1,'--o');
xlabel('Job #');
ylabel('Number of examples in the proximity');

figure;
plot(cycletimeforecastingexample(151:200,1:1),lda1,'k--o');
xlabel('Job #');
ylabel('Local Prediction Accuracy (RMSE) (hrs)');

alpha1=zeros(max(size(test_x)),1);
beta1=zeros(max(size(test_x)),1);
theta1=zeros(max(size(test_x)),1);

lo1=0.0105;
lo3=0.15;

alpha1(:,1)=exp(-lo1*train_rmse*ones(max(size(test_x)),1));
beta1(:,1)=1-alpha1(:,1);
theta1=1-exp(-lo3*lda1./freq1);

figure;
plot(cycletimeforecastingexample(151:200,1:1),alpha1,cycletimeforecastingexample(151:200,1:1),b
eta1,'--',cycletimeforecastingexample(151:200,1:1),theta1,'b-*');
title('Three Memberships of Jobs in Test Data');
xlabel('Job #');
ylabel('Membership');
legend('Truth membership','Falsity membership','Indeterminacy membership');
```

Fig. 2.7 MATLAB code for evaluating the neighborhood crowding degree and local prediction accuracy of each job in test data

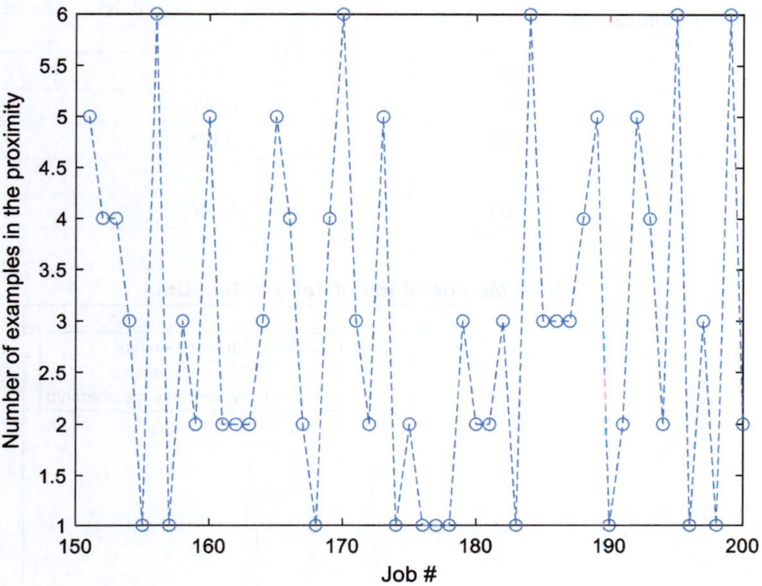

Fig. 2.8 Neighborhood crowding degree evaluation results

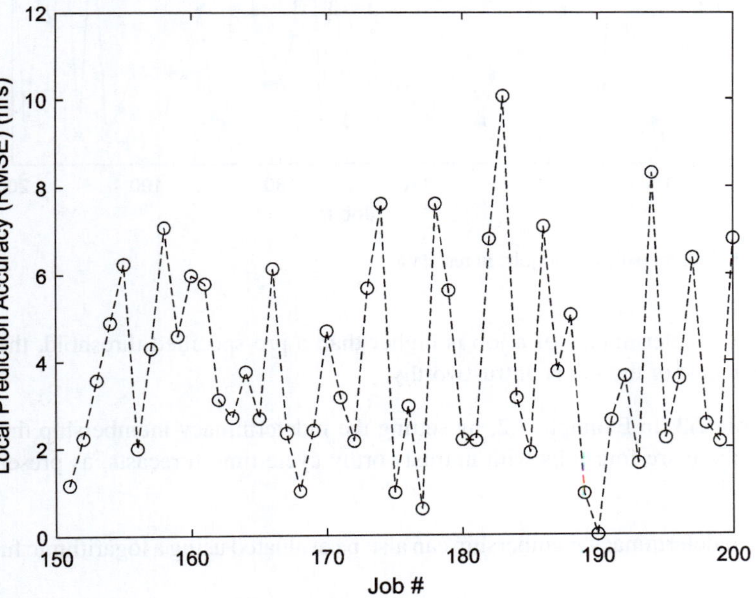

Fig. 2.9 Local prediction accuracy around each job

Table 2.3 Indeterminacy of jobs in test data

j	$\theta_{\tilde{y}_j}$
151	0.034
152	0.080
153	0.125
…	
200	0.400

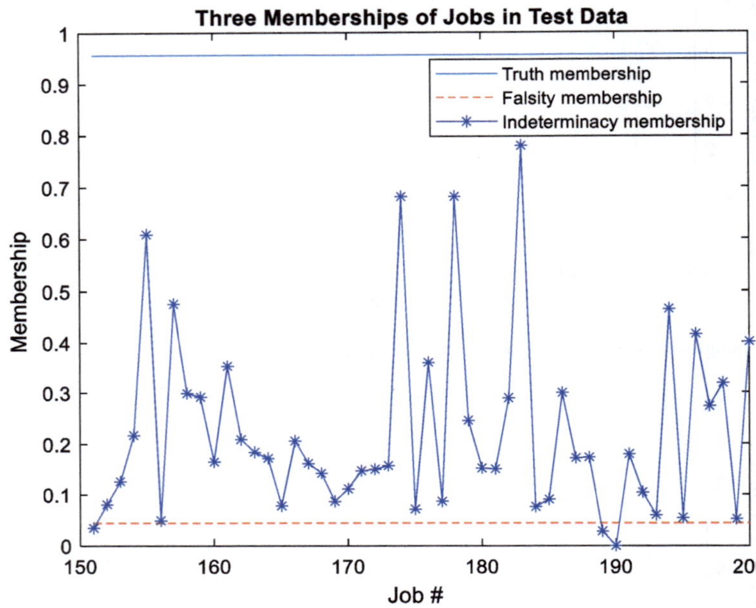

Fig. 2.10 Memberships of all jobs in test data

If the indeterminacy of a job is higher than a pre-specified threshold, the cycle time forecast of the job is untrustworthy.

Example 2.3 In Example 2.2, by setting the indeterminacy membership threshold to 0.5, there are four jobs with untrustworthy cycle time forecasts, as presented in Table 2.4.

The indeterminacy membership can also be evaluated using a logarithmic function as

$$\theta_{\tilde{y}_j} = 1 - \log\left(\frac{1}{\rho_3 \cdot \frac{RMSE(prox(j))}{|prox(j)|} + c} + 1\right) \tag{2.25}$$

Table 2.4 Jobs with untrustworthy cycle time forecasts

j	$\theta_{\tilde{y}_j}$
155	0.608
174	0.681
178	0.681
183	0.779

The neutrosophic forecasting approach introduced in this section is a general methodology, and can be applied to other forecasting problems, including time-series forecasting problems, as described below.

Example 2.4 Chen and Lin [12] proposed a neutrosophic classification method for trustworthy predictive maintenance in smart manufacturing, in which a DNN is constructed and trained to predict the next time to failure of a machine (y_j) based on three sensor readings for monitoring vibration (acceleration), temperature, and frequency, so as to arrange preventive maintenance in advance [14–16]. The forecasting accuracy for the training data is $RMSE = 3.43$ (hr). Consider the first case in test data for fold #4. $y_1 = 401.8$ (hr). Set $\in$, ρ_3, and ρ_4 to 20, 3.5, and 0.5, respectively:

$$\alpha_{\tilde{y}_1} = 3.5 \cdot \ln\left(\frac{1}{3.43} + 1\right) = 0.895$$

$$\beta_{\tilde{y}_1} = 1 - 0.895 = 0.105$$

Five examples in the training data are in the proximity of this case, so $|prox(\mathbf{x}_1)| = 5$. The local classification accuracy in terms of $RMSE(prox(\mathbf{x}_1)) = 3.21$ (hr). Therefore,

$$\theta_{\tilde{y}_1} = 1 - 5 \cdot \ln\left(\frac{1}{\frac{3.21}{5}} + 1\right) = 0.530$$

Therefore, the neutrosophic time-to-failure forecast of this case is 401.8; 0.895, 0.530, 0.105, in this example, with the indeterminacy threshold ψ set at 0.5. Therefore, the time-to-failure forecast of the case is not trustworthy and should not be casually communicated to the engineer. In addition, among the 300 cases in test data, the time-to-failure prediction results for 17 cases are untrustworthy and should not be casually communicated to the engineer. This conclusion is irrelevant to whether these prediction results were ultimately proven accurate.

2.4.2 Time-Series Forecasting Problems

In a time-series forecasting problem, inputs to the forecasting model are the values of the target variable during the past few periods. For simple time-series forecasting problems, moving average (MA) [5] and exponential smoothing (ES) [6] are two prevalent forecasting methods:

(Ma)

$$\hat{y}_t = \frac{\sum_{l=1}^{n-1} y_{t-l}}{n} = \sum_{l=1}^{n-1} \frac{1}{n} \cdot y_{t-l} \tag{2.26}$$

(Es)

$$\hat{y}_t = \alpha y_{t-1} + (1-\alpha)\hat{y}_{t-1} = \sum_{l=1}^{t-1} \alpha(1-\alpha)^{l-1} \cdot y_{t-l} + (1-\alpha)^{t-1} \cdot y_0 \tag{2.27}$$

Both are in the form of weighted average and can take indeterminacy into account through applying NWA. NWA is directly applicable if historical data can be expressed as TNNs. The systematic mechanism introduced in Sect. 2.3.1 can be applied to derive the indeterminacy of the forecasting result.

A more sophisticated time-series forecasting method is autoregressive integrated moving average (ARIMA). In the previous example, the cycle time of a job can also be predicted from the cycle times of the previous p jobs using [17]

$$\hat{y}_j = ARIMA(\{y_{j-k} | k = 1 \sim p\}\ ; p, d, q);\ j > n \tag{2.28}$$

satisfying

$$\left(1 - \sum_{k=1}^{p} \varphi_k L^k\right)(1-L)^d y_j = \delta + \left(1 + \sum_{k=1}^{q} \theta_k L^k\right)\varepsilon_j \tag{2.29}$$

where L is the lag operator; $\{\varphi_k\}$ are the parameters of the autoregressive part of the model; $\{\theta_k\}$ are the parameters of the moving average part; ε_j is an error term.

The same formulae mentioned above are applicable to evaluating the membership functions. However, the distance between two jobs is measured as

$$dist(l, j) = \sqrt{\sum_{k=1}^{p} (y_{l-k} - y_{j-k})^2} \tag{2.30}$$

Example 2.5 The cycle times of 200 jobs for the same product type have been collected, as presented in Table 2.5. The data of the first 150 jobs are used as the training data, while the remaining data are reserved for testing. ARIMA is applied to

Table 2.5 Cycle time series of 200 jobs

j	y_j (hrs)
1	677
2	687
3	687
…	
200	1373

predict the cycle times of jobs in test data, for which the MATLAB code is provided in Fig. 2.11. The optimal value of parameters $p^* = 3$, $d^* = 2$, and $q^* = 4$. Then, the cycle time forecasting performance is evaluated as RMSE = 20.7 (hrs). ρ_1 is set to 0.0105. Therefore, the truth and falsity memberships can be evaluated as.

$$\alpha_{\hat{y}_j} = e^{-0.0105 \cdot 20.7} = 0.805$$

$$\beta_{\hat{y}_j} = 1 - 0.805 = 0.195$$

To evaluate the indeterminacy membership, $\in$ is set to 650. The neighborhood crowding degree of each job in test data is evaluated and shown in Fig. 2.12. The local prediction accuracy around each job is evaluated and shown in Fig. 2.13. By setting $\rho_3 = 0.025$, the indeterminacy membership of each job in test data can be evaluated, as shown in Fig. 2.14.

2.5 Collaboration Among Neutrosophic Forecasting Methods

Multiple analyses of a forecasting problem from diverse perspectives raise the chance that no relevant aspects of the forecasting problem will be ignored. Assume Q neutrosophic forecasting models (or methods) are applied to predict the cycle time of job j, and the results are indicated with $\hat{y}_j(q)$; $q = 1 \sim Q$, respectively. These single-valued neutrosophic forecasts can be simply aggregated using weighted average (WA) as

$$\hat{y}_j = \frac{\sum_{q=1}^{Q} \left(w_j \cdot \hat{y}_j(q) \right)}{\sum_{q=1}^{Q} w_j} \tag{2.31}$$

Then, according to the logic operations for single-valued neutrosophic sets [7, 8],

$$\alpha_{\hat{y}_j} = \min_q \alpha_{\hat{y}_j(q)} \tag{2.32}$$

```
cd F:\ACTM
example1=readtable("example1.xlsx");
example1=table2array(example1);
train_y=example1(1:150, 2:2);
test_y=example1(151:200, 2:2);

d=0;
data1=train_y;
if adftest(train_y)==0 || kpsstest(train_y)==1
  dtrain_y=diff(train_y);
  data1= dtrain_y;
  d=1;
end
if adftest(dtrain_y)==0 || kpsstest(dtrain_y)==1
  ddtrain_y=diff(dtrain_y);
  data1= ddtrain_y;
  d=2;
end

data1=reshape(data1, length(data1), 1);
p_max=4;
q_max=4;
logL_all=zeros(p_max+1, q_max+1);
p_and_q=zeros(p_max+1, q_max+1);
for p=0:p_max
  for q=0:q_max
    model=arima(p, d, q);
    [fit, ~, logL]=estimate(model, data1);
    logL_all(p+1, q+1)=logL;
    p_and_q(p+1, q+1)=p+q;
  end
end
logL_all=reshape(logL_all, (p_max+1)*(q_max+1), 1);
p_and_q=reshape(p_and_q, (p_max+1)*(q_max+1), 1);
m2=length(data1);
[aic, bic]=aicbic(logL_all, p_and_q+1, m2);
aic0=reshape(aic, (p_max+1), (q_max+1));
bic0=reshape(bic, (p_max+1), (q_max+1));

aic1=min(aic0(:));
index1=(aic1==aic0);
[p_all q_all]=meshgrid(0:p_max, 0:q_max);
p0=p_all(index1);
q0=q_all(index1);

aic2=min(bic0(:));
index1=(aic2==bic0);
[p_all q_all]=meshgrid(0:p_max, 0:q_max);
p1=p_all(index1);
q1=q_all(index1);

if p0^2+q0^2>p1^2+q1^2
  p=p1;
  q=q1;
else
  p=p0;
  q=q0;
end

Mdl=arima(p, d, q);
fit1=estimate(Mdl, data1);
[test_o, var_test_o]=forecast(fit1, 50, 'Y0', train_y);
test_rmse=mean((test_y-test_o).^2)^0.5;
```

Fig. 2.11 MATLAB code for implementing ARIMA

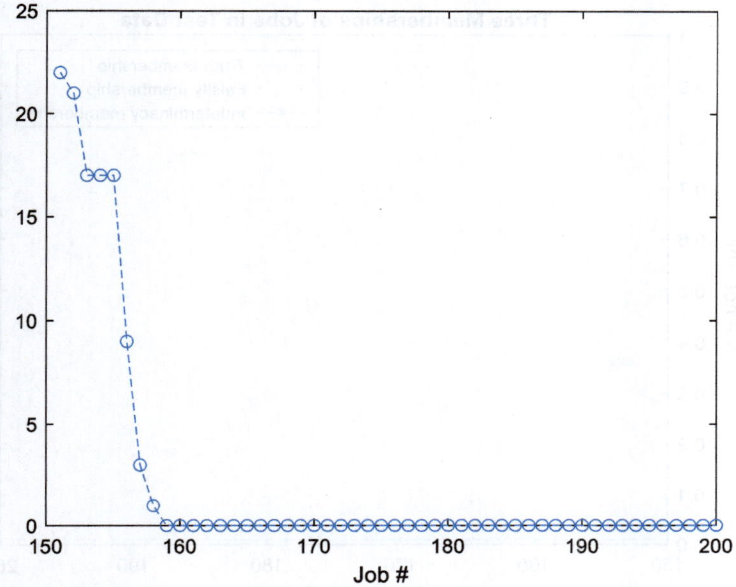

Fig. 2.12 Neighborhood crowding degree of each job in test data in cycle time-series forecasting

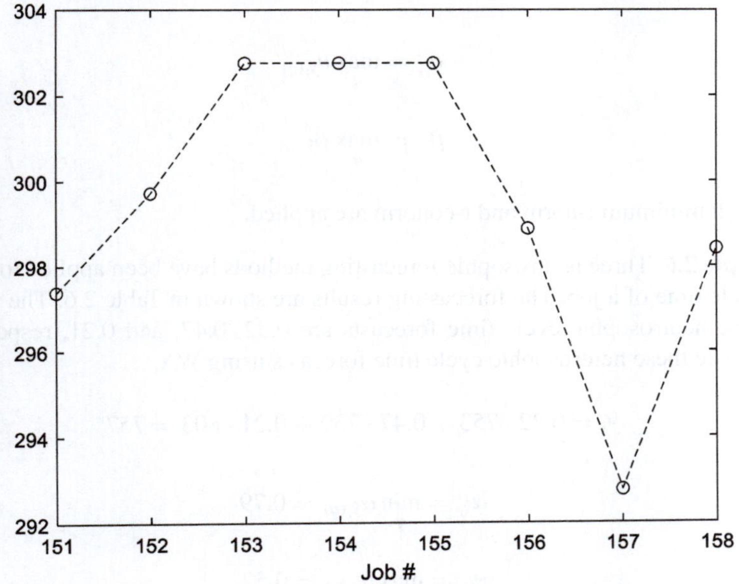

Fig. 2.13 Local prediction accuracy around each job in cycle time-series forecasting

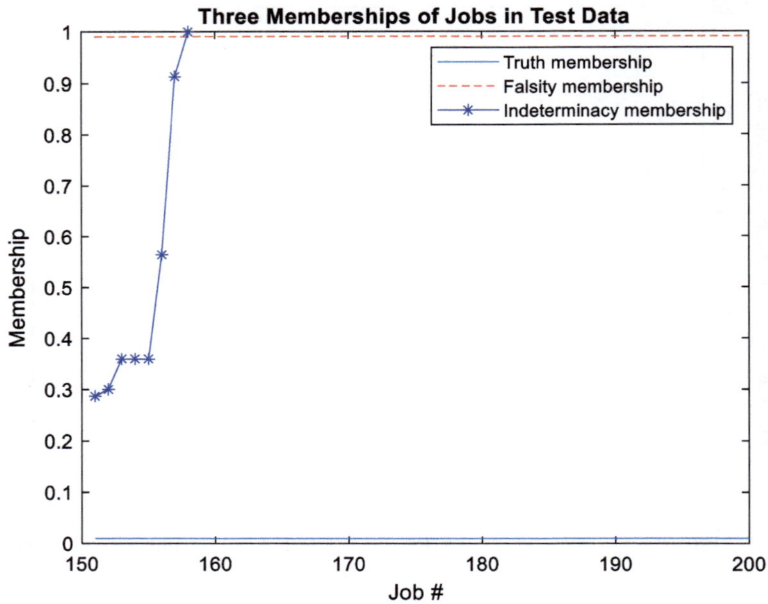

Fig. 2.14 Memberships of cycle time forecasts for jobs in test data in cycle time-series forecasting

$$\theta_{\hat{y}_j} = \max_q \theta_{\hat{y}_j(q)} \tag{2.33}$$

$$\beta_{\hat{y}_j} = \max_q \beta_{\hat{y}_j(q)} \tag{2.34}$$

in which minimum t-norm and t-conorm are applied.

Example 2.6 Three neutrosophic forecasting methods have been applied to predict the cycle time of a job. The forecasting results are shown in Table 2.6. The weights of these neutrosophic cycle time forecasts are 0.32, 0.47, and 0.21, respectively. Aggregate these neutrosophic cycle time forecasts using WA,

$$\hat{y}_j = 0.32 \cdot 753 + 0.47 \cdot 739 + 0.21 \cdot 803 = 757$$

$$\alpha_{\hat{y}_j} = \min_q \alpha_{\hat{y}_j(q)} = 0.79$$

$$\theta_{\hat{y}_j} = \max_q \theta_{\hat{y}_j(q)} = 0.52$$

$$\beta_{\hat{y}_j} = \max_q \beta_{\hat{y}_j(q)} = 0.21$$

Table 2.6 Single-valued cycle time forecasts for job j

q	$\hat{y}_j(q)$(hrs)
1	753; 0.88, 0.42, 0.12
2	739; 0.82, 0.47, 0.18
3	803; 0.79, 0.52, 0.21

Therefore, $\hat{y}_j = 757; 0.79, 0.52, 0.21$.

If Archimedean t-norm and t-conorm are applied instead [1],

$$\alpha_{\hat{y}_j} = \xi_Q\left(\left\{\alpha_{\hat{y}_j(q)}\right\}\right) \tag{2.35}$$

$$\theta_{\hat{y}_j} = \prod_{q=1}^{Q} \theta_{\hat{y}_j(q)} \tag{2.36}$$

$$\beta_{\hat{y}_j} = \prod_{q=1}^{Q} \beta_{\hat{y}_j(q)} \tag{2.37}$$

where

$$\xi_c\left(\left\{\alpha_{\hat{y}_j(q)}\right\}\right) = \begin{cases} 0 & if \quad c = 0 \\ \alpha_{\hat{y}_j(c)} + \xi_{c-1}\left(\left\{\alpha_{\hat{y}_j(q)}\right\}\right) - \alpha_{\hat{y}_j(c)}\xi_{c-1}\left(\left\{\alpha_{\hat{y}_j(q)}\right\}\right) & otherwise \end{cases} \tag{2.38}$$

Example 2.7 In the previous example, Archimedean t-norm and t-conorm are applied to aggregate the neutrosophic cycle time forecasts using WA as.

$$\xi_0\left(\left\{\alpha_{\hat{y}_j(q)}\right\}\right) = 0$$

$$\begin{aligned} \xi_1\left(\left\{\alpha_{\hat{y}_j(q)}\right\}\right) &= \alpha_{\hat{y}_j(1)} + \xi_0\left(\left\{\alpha_{\hat{y}_j(q)}\right\}\right) \\ &\quad - \alpha_{\hat{y}_j(1)}\xi_0\left(\left\{\alpha_{\hat{y}_j(q)}\right\}\right) \\ &= \alpha_{\hat{y}_j(1)} = 0.88 \end{aligned}$$

$$\begin{aligned} \xi_2\left(\left\{\alpha_{\hat{y}_j(q)}\right\}\right) &= \alpha_{\hat{y}_j(2)} + \xi_1\left(\left\{\alpha_{\hat{y}_j(q)}\right\}\right) \\ &\quad - \alpha_{\hat{y}_j(2)}\xi_1\left(\left\{\alpha_{\hat{y}_j(q)}\right\}\right) \\ &= 0.82 + 0.88 - 0.82 \cdot 0.88 \\ &= 0.978 \end{aligned}$$

$$\begin{aligned} \xi_3\left(\left\{\alpha_{\hat{y}_j(q)}\right\}\right) &= \alpha_{\hat{y}_j(3)} + \xi_2\left(\left\{\alpha_{\hat{y}_j(q)}\right\}\right) \\ &\quad - \alpha_{\hat{y}_j(3)}\xi_2\left(\left\{\alpha_{\hat{y}_j(q)}\right\}\right) \end{aligned}$$

$$= 0.79 + 0.978 - 0.79 \cdot 0.978$$
$$= 0.995$$

$$\alpha_{\hat{y}_j} = \xi_3\left(\left\{\alpha_{\hat{y}_j(q)}\right\}\right) = 0.995$$

$$\theta_{\hat{y}_j} = 0.42 \cdot 0.47 \cdot 0.52 = 0.103$$

$$\beta_{\hat{y}_j} = 0.12 \cdot 0.18 \cdot 0.21 = 0.005$$

Therefore, $\langle \hat{y}_j = 757; 0.995, 0.103, 0.005 \rangle$.

2.6 Neutrosophic Forecasting Method with Interval Forecasts

The range of y_j can be estimated by generating an interval-valued or fuzzy forecast. For example, the output node threshold in Eq. (2.16), θ^o, can be fuzzified to generate a fuzzy forecast so that actual value (y_j) belongs to the corresponding fuzzy forecast ($\tilde{y}_j$) [18]:

$$\tilde{y}_j = \sum_{l=1}^{L} w_l^o h_{jl} - \tilde{\theta}^o \tag{2.39}$$

such that

$$y_{j1} = \sum_{l=1}^{L} w_l^o h_{jl} - \theta_3^o \leq y_j \tag{2.40}$$

$$y_{j2} = \sum_{l=1}^{L} w_l^o h_{jl} - \theta_2^o = \hat{y}_j \tag{2.41}$$

$$y_{j3} = \sum_{l=1}^{L} w_l^o h_{jl} - \theta_1^o \geq y_j \tag{2.42}$$

Substituting Eq. (2.41) into Eq. (2.40) gives

$$y_{j2} + \theta_2^o - \theta_3^o \leq y_j \tag{2.43}$$

Therefore,

$$\theta_3^o \geq y_{j2} - y_j + \theta_2^o \tag{2.44}$$

To minimize the range of y_j [18],

$$\theta_3^o = \max_j(y_{j2} - y_j) + \theta_2^o \tag{2.45}$$

Substituting Eq. (2.45) into Eq. (2.40) gives

$$y_{j1} = \sum_{l=1}^{L} w_l^o h_{jl} - \left(\max_r(y_{r2} - y_r) + \theta_2^o\right) = y_{j2} - \max_r(y_{r2} - y_r) \tag{2.46}$$

Similarly,

$$y_{j3} = y_{j2} - \min_r(y_{r2} - y_r) \tag{2.47}$$

Example 2.8 In Example 2.1, the output node threshold is fuzzified to generate a fuzzy forecast so that actual value belongs to the corresponding fuzzy forecast. The required MATLAB code is provided in Fig. 2.15. The estimation results are summarized in Fig. 2.16.

Another ANN parameter that can be fuzzified for similar purposes is w_l^o. Consider w_ς^o,

$$\tilde{y}_j = \tilde{w}_\varsigma^o h_{j\varsigma} + \sum_{l \neq \varsigma} w_l^o h_{jl} - \theta^o \tag{2.48}$$

If $h_{j\varsigma} \geq 0$,

$$y_{j1} = w_{\varsigma 1}^o h_{j\varsigma} + \sum_{l \neq \varsigma} w_l^o h_{jl} - \theta^o \leq a_j \tag{2.49}$$

```
UB=test_o+max(train_y-train_o);
LB=test_o+min(train_y-train_o);
plot(cycletimeforecastingexample(151:200,1:1),test_y,'o',cycletimeforecastingexample(151:200,1:1)
,UB,'--
',cycletimeforecastingexample(151:200,1:1),test_o,cycletimeforecastingexample(151:200,1:1),LB,'k
--');
title('Fuzzy Cycle Time Forecasts');
xlabel('Job #');
ylabel('Cycle time (hrs)');
legend('Actual value','Upper bound','Core','Lower bound');
ylim([800 2000]);
```

Fig. 2.15 MATLAB code for generating fuzzy cycle time forecasts

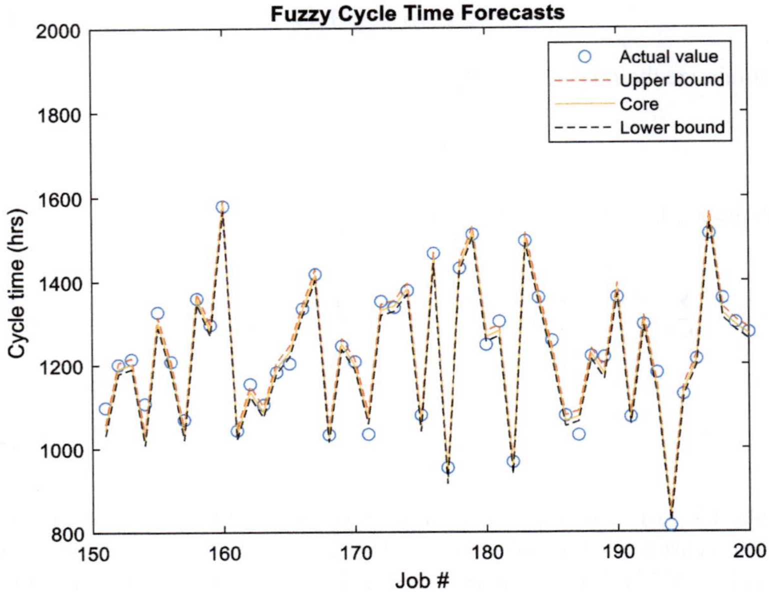

Fig. 2.16 Fuzzy cycle time forecasts

$$y_{j3} = w^o_{\varsigma 3} h_{j\varsigma} + \sum_{l \neq \varsigma} w^o_l h_{jl} - \theta^o \geq a_j \tag{2.50}$$

Therefore,

$$w^o_{\varsigma 1} \leq \frac{a_j - \sum_{l \neq \varsigma} w^o_l h_{jl} + \theta^o}{h_{j\varsigma}} \tag{2.51}$$

$$w^o_{\varsigma 3} \geq \frac{a_j - \sum_{l \neq \varsigma} w^o_l h_{jl} + \theta^o}{h_{j\varsigma}} \tag{2.52}$$

To optimize the forecasting precision,

$$w^{o*}_{\varsigma 1} = \min_j \left(\frac{a_j - \sum_{l \neq \varsigma} w^o_l h_{jl} + \theta^o}{h_{j\varsigma}} \right) \tag{2.53}$$

$$w^{o*}_{\varsigma 3} = \max_j \left(\frac{a_j - \sum_{l \neq \varsigma} w^o_l h_{jl} + \theta^o}{h_{j\varsigma}} \right) \tag{2.54}$$

Substituting Eqs. (2.53) and (2.54) into Eqs. (2.39) and (2.41) gives

$$y_{j1} = \min_r\left(\frac{a_r - \sum_{l \neq \varsigma} w_l^o h_{rl} + \theta^o}{h_{r\varsigma}}\right) h_{j\varsigma} + \sum_{l \neq \varsigma} w_l^o h_{jl} - \theta^o \tag{2.55}$$

$$y_{j3} = \max_r\left(\frac{a_r - \sum_{l \neq \varsigma} w_l^o h_{rl} + \theta^o}{h_{r\varsigma}}\right) h_{j\varsigma} + \sum_{l \neq \varsigma} w_l^o h_{jl} - \theta^o \tag{2.56}$$

Conversely, if $h_{j\varsigma} < 0$,

$$y_{j1} = w_{\varsigma 3}^o h_{j\varsigma} + \sum_{l \neq \varsigma} w_l^o h_{jl} - \theta^o \leq a_j \tag{2.57}$$

$$y_{j3} = w_{\varsigma 1}^o h_{j\varsigma} + \sum_{l \neq \varsigma} w_l^o h_{jl} - \theta^o \geq a_j \tag{2.58}$$

Therefore,

$$w_{\varsigma 3}^o \geq \frac{a_j - \sum_{l \neq \varsigma} w_l^o h_{jl} + \theta^o}{h_{j\varsigma}} \tag{2.59}$$

$$w_{\varsigma 1}^o \leq \frac{a_j - \sum_{l \neq \varsigma} w_l^o h_{jl} + \theta^o}{h_{j\varsigma}} \tag{2.60}$$

To optimize the forecasting precision,

$$w_{\varsigma 3}^{o*} = \max_j\left(\frac{a_j - \sum_{l \neq \varsigma} w_l^o h_{jl} + \theta^o}{h_{j\varsigma}}\right) \tag{2.61}$$

$$w_{\varsigma 1}^{o*} = \min_j\left(\frac{a_j - \sum_{l \neq \varsigma} w_l^o h_{jl} + \theta^o}{h_{j\varsigma}}\right) \tag{2.62}$$

which are exactly the same as Eqs. (2.53) and (2.54).

However, since the tanh activation function confines the output from each hidden layer to be within [–1, 1], fuzzifying other parameters cannot guarantee that all fuzzy forecasts contain the corresponding actual values [18].

Example 2.9 In Example 2.1, the connection weight between the output node and the first hidden-layer node is fuzzified to generate a fuzzy forecast so that each fuzzy cycle time forecast contains the corresponding actual value. The estimation results are summarized in Fig. 2.17.

The maximum or minimum of the three memberships of each fuzzy forecast can be derived using Eqs. (2.19), (2.20), and (2.23). Then, if triangular neutrosophic numbers (TNNs) are adopted,

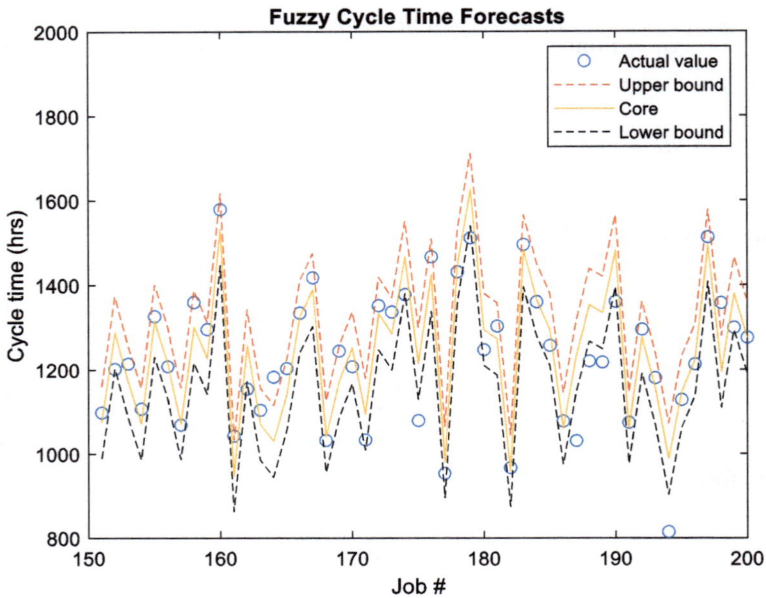

Fig. 2.17 Fuzzy cycle time forecasts by fuzzifying the connection weight

$$T_{\tilde{y}_j}(x) = \begin{cases} \alpha_{\tilde{y}_j}\left(\frac{x-y_{j1}}{y_{j2}-y_{j1}}\right) & if \quad y_{j1} \le x < y_{j2} \\ \alpha_{\tilde{y}_j}\left(\frac{y_{j3}-x}{y_{j3}-y_{j2}}\right) & if \quad y_{j2} \le x < y_{j3} \\ 0 & otherwise \end{cases} \tag{2.63}$$

$$I_{\tilde{y}_j}(x) = \begin{cases} \frac{y_{j2}-x+\theta_{\tilde{y}_j}(x-y_{j1})}{y_{j2}-y_{j1}} & if \quad y_{j1} \le x < y_{j2} \\ \frac{x-y_{j2}+\theta_{\tilde{y}_j}(y_{j3}-x)}{y_{j3}-y_{j2}} & if \quad y_{j2} \le x < y_{j3} \\ 0 & otherwise \end{cases} \tag{2.64}$$

$$F_{\tilde{y}_j}(x) = \begin{cases} \frac{y_{j2}-x+\beta_{\tilde{y}_j}(x-y_{j1})}{y_{j2}-y_{j1}} & if \quad y_{j1} \le x < y_{j2} \\ \frac{x-y_{j2}+\beta_{\tilde{y}_j}(y_{j3}-x)}{y_{j3}-y_{j2}} & if \quad y_{j2} \le x < y_{j3} \\ 0 & otherwise \end{cases} \tag{2.65}$$

The forecast is indicated with $\hat{y}_j = y_{j1}, y_{j2}, y_{j3}; \alpha_{\tilde{y}_j}, \theta_{\tilde{y}_j}, \beta_{\tilde{y}_j}$. An example is provided in Fig. 2.18, where $\hat{y}_j = 902, 1026, 1235; 0.8, 0.4, 0.2$.

There are various methods for aggregating forecasts expressed in TNNs. For example, NWA

$$\hat{y}_j = NWA(\hat{y}_j(q)) = y_{j1}, y_{j2}, y_{j3}; \alpha_{\tilde{y}_j}, \theta_{\tilde{y}_j}, \beta_{\tilde{y}_j} \tag{2.66}$$

where

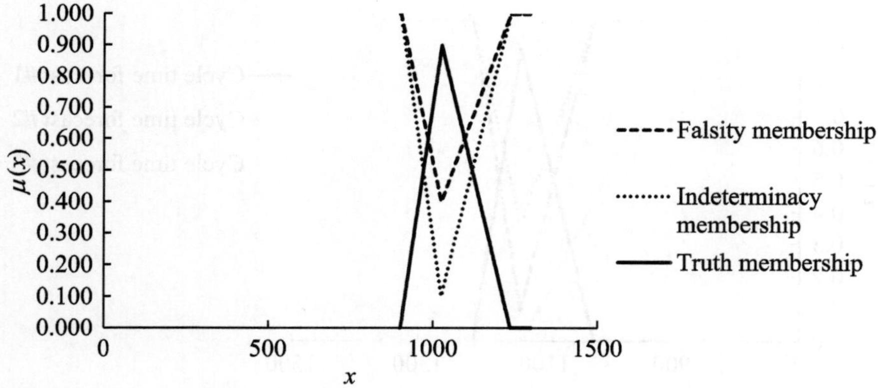

Fig. 2.18 Triangular neutrosophic number

$$y_{j1} = \frac{\sum_{q=1}^{Q} \left(w_j \cdot \hat{y}_{j1}(q)\right)}{\sum_{q=1}^{Q} w_j} \tag{2.67}$$

$$y_{j2} = \frac{\sum_{q=1}^{Q} \left(w_j \cdot \hat{y}_{j2}(q)\right)}{\sum_{q=1}^{Q} w_j} \tag{2.68}$$

$$y_{j3} = \frac{\sum_{q=1}^{Q} \left(w_j \cdot \hat{y}_{j3}(q)\right)}{\sum_{q=1}^{Q} w_j} \tag{2.69}$$

$$\alpha_{\hat{y}_j} = \min_q \alpha_{\hat{y}_j(q)} \tag{2.70}$$

$$\theta_{\hat{y}_j} = \max_q \theta_{\hat{y}_j(q)} \tag{2.71}$$

$$\beta_{\hat{y}_j} = \max_q \beta_{\hat{y}_j(q)} \tag{2.72}$$

Example 2.10 Three cycle time forecasts in TNNs are shown in Fig. 2.19:

$$\hat{y}_j(1) = 902, 1026, 1235; 0.9, 0.4, 0.1$$

$$\hat{y}_j(2) = 852, 914, 1135; 0.86, 0.45, 0.14$$

$$\hat{y}_j(3) = 1023, 1139, 1214; 0.91, 0.36, 0.09$$

Their NWI result is provided in Fig. 2.20, for which their weights are set to 0.3, 0.2, and 0.5, respectively.

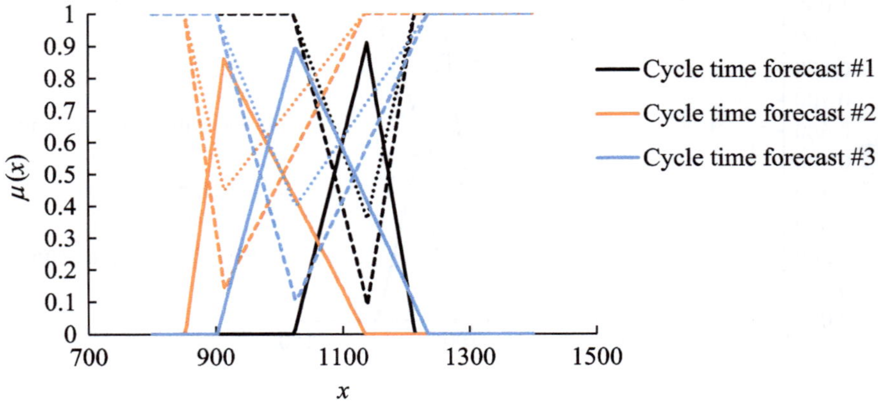

Fig. 2.19 Three forecasts in TNNs

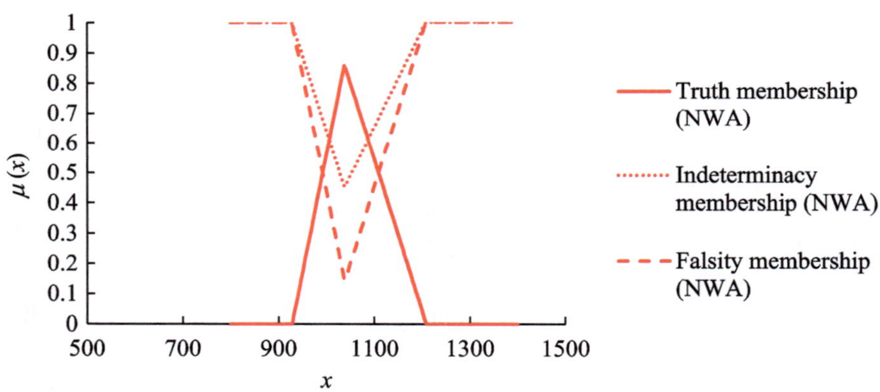

Fig. 2.20 Aggregation result using NWI

NWI is another prevalent aggregator for multiple neutrosophic forecasts with unequal weights [1]. The NWI operator is an extension of the FWI operator [18–21]. In NWI, the aggregation result of M neutrosophic forecasts, $\{\hat{y}_j(m), m = 1 \sim M\}$, is aggregated as $\hat{y}_j$, such that

$$\mu_{\hat{y}_j}(x) = \min_m \mu_{\hat{y}_j(m)}(x) + \sum_m (\omega_m - \min_r \omega_r)(\mu_{\hat{y}_j(m)}(x) - \min_r \mu_{\hat{y}_j(r)}(x)) \quad (2.73)$$

The membership functions of $\hat{y}_j$ are

$$\alpha_{\hat{y}_j} = \max(\min_m \alpha_{\hat{y}_j(m)}(x) + \sum_m (\omega_m - \min_r \omega_r)(\alpha_{\hat{y}_j(m)}(x) - \min_r \alpha_{\hat{y}_j(r)}(x))) \quad (2.74)$$

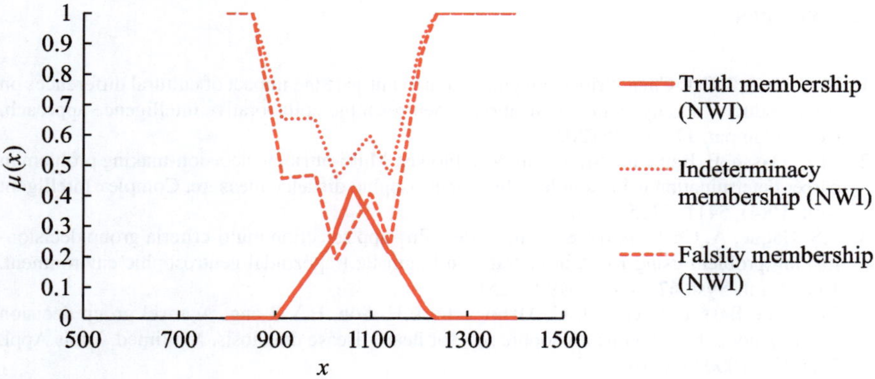

Fig. 2.21 NWI result of the three neutrosophic forecasts in Example 2.9

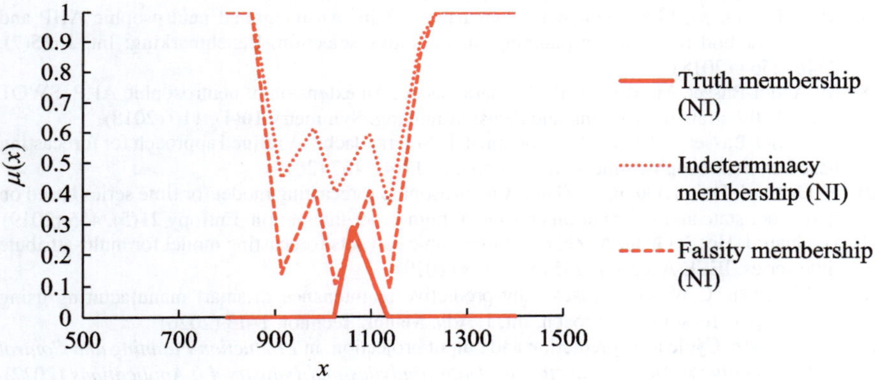

Fig. 2.22 NI result of the three neutrosophic forecasts in Example 2.9

$$\theta_{\hat{y}_j} = \min(\min_m \theta_{\hat{y}_j(m)}(x) + \sum_m (\omega_m - \min_r \omega_r)(\theta_{\hat{y}_j(m)}(x) - \min_r \theta_{\hat{y}_j(r)}(x))) \tag{2.75}$$

$$\beta_{\hat{y}_j} = \min(\min_m \beta_{\hat{y}_j(m)}(x) + \sum_m (\omega_m - \min_r \omega_r)(\beta_{\hat{y}_j(m)}(x) - \min_r \beta_{\hat{y}_j(r)}(x))) \tag{2.76}$$

Example 2.11 In the previous example, NWI is applied to aggregate the three forecasts instead. Figure 2.21 depicts the NWI result.

When the weights of neutrosophic forecasts are equal, NWI reduces to NI.

Example 2.12 In the previous example, NI is applied to aggregate the three forecasts instead. Figure 2.22 depicts the NI result.

References

1. M.C. Chiu, T.C.T. Chen, Prioritizing measures to mitigate the impact of cultural differences on semiconductor supply chain localization: a neutrosophic collaborative intelligence approach. Cogn. Comput. **17**(4), 125 (2025)
2. M.A. Dasan, E. Bementa, M. Aslam, V.L. Flower, Multi-attribute decision-making problem in career determination using single-valued neutrosophic distance measure. Complex Intelligent Syst. **10**(4), 5411–5425 (2024)
3. T.S. Haque, A. Chakraborty, S. Alam, E-learning app selection multi-criteria group decision-making problem using Einstein operator in linguistic trapezoidal neutrosophic environment. Knowl. Inf. Syst. **67**, 2481–2519 (2025)
4. M. Abdel-Basset, A. Gamal, G. Manogaran, L.H. Son, H.V. Long, A novel group decision making model based on neutrosophic sets for heart disease diagnosis. Multimed. Tools Appl. **79**, 9977–10002 (2020)
5. S. Hansun, A new approach of moving average method in time series analysis. in *2013 Conference on New Media Studies* (2013), pp. 1–4
6. E. Ostertagova, O. Ostertag, Forecasting using simple exponential smoothing method. Acta Electrotechnica et Informatica **12**(3), 62 (2012)
7. M.A. Basset, M. Mohamed, A.K. Sangaiah, V. Jain, An integrated neutrosophic AHP and SWOT method for strategic planning methodology selection. Benchmarking: Int. J. **25**(7), 2546–2564 (2018)
8. M. Abdel-Basset, M. Mohamed, F. Smarandache, An extension of neutrosophic AHP–SWOT analysis for strategic planning and decision-making. Symmetry **10**(4), 116 (2018)
9. M. Abdel-Basset, V. Chang, M. Mohamed, F. Smarandache, A refined approach for forecasting based on neutrosophic time series. Symmetry **11**(4), 457 (2019)
10. H. Guan, Z. Dai, S. Guan, A. Zhao, A neutrosophic forecasting model for time series based on first-order state and information entropy of high-order fluctuation. Entropy **21**(5), 455 (2019)
11. H. Guan, J. He, S. Guan, A. Zhao, Neutrosophic soft sets forecasting model for multi-attribute time series. IEEE Access **7**, 25575–25588 (2019)
12. T.C.T. Chen, C.W. Lin, Trustworthy predictive maintenance in smart manufacturing using neutrosophic reasoning approach. Int. J. Adv. Manuf. Technol. 1–13 (2026)
13. T.C.T. Chen, Cycle time prediction and output projection. in *Production Planning and Control in Semiconductor Manufacturing: Big Data Analytics and Industry 4.0 Applications* (2022), pp. 41–62
14. S. Namuduri, B.N. Narayanan, V.S.P. Davuluru, L. Burton, S. Bhansali, Deep learning methods for sensor based predictive maintenance and future perspectives for electrochemical sensors. J. Electrochem. Soc. **167**(3), 037552 (2020)
15. T.C.T. Chen, M.C. Chiu, H.C. Wu, Fuzzy-neural cloud-edge computing application for internal due date assignment: wafer fabrication as an example. Prog. Artif. Intell. 1–18 (2026)
16. M. Pech, J. Vrchota, J. Bednář, Predictive maintenance and intelligent sensors in smart factory. Sensors **21**(4), 1470 (2021)
17. D.M. Xue, Z.Q. Hua, ARIMA based time series forecasting model. Recent Advances in Electrical & Electronic Engineering (Formerly Recent Patents on Electrical & Electronic Engineering) **9**(2), 93–98 (2016)
18. T.C.T. Chen, Y.C. Wang, C.W. Lin, A fuzzy collaborative forecasting approach considering experts' unequal levels of authority. Appl. Soft Comput. **94**, 106455 (2020)
19. T.-C.T. Chen, C.-W. Lin, Y.-C. Wang, An auto-weighting FWI fuzzy collaborative intelligence approach for forecasting DRAM yield. Procedia Manufact. **55**, 102–109 (2021)
20. T. Chen, Y.-C. Wang, H.-C. Wu, Analyzing the impact of vaccine availability on alternative supplier selection amid the COVID-19 pandemic: a cFGM-FTOPSIS-FWI approach. Healthcare **9**(1), 71 (2021)
21. T. Chen, H.-C. Wu, A partial-consensus and unequal-authority fuzzy collaborative intelligence approach for assessing robotic applications amid the COVID-19 pandemic. Soft. Comput. **27**, 16493–16509 (2023)

Chapter 3
Neutrosophic Collaborative Decision-Making

Abstract This chapter begins by reviewing existing literature on neutrosophic decision-making methods, then focuses on the universality of embedding neutrosophic logic into the analytic hierarchy process (AHP)—the technique for order preference by similarity to an ideal solution (TOPSIS). Next, the chapter describes the process of neutrosophic collaborative decision-making. This process begins with decision makers expressing their neutrosophic judgments on the relative weights of various criteria. Based on the neutrosophic judgment matrix, various methods can be applied to derive the criterion weights, such as neutrosophic geometric mean (NGM), neutrosophic extent analysis (NEA), deneutrosophication + arithmetic mean (AR), deneutrosophication + eigen analysis, and calibrated NGM (cNGM). Finally, this chapter introduces the neutrosophic TOPSIS and its applications. The decision results of multiple decision makers can be aggregated using techniques such as neutrosophic weighted average and neutrosophic weighted intersection. However, aggregating the neutrosophic judgments of multiple decision makers, the indeterminacy of the aggregation result is usually the highest. To resolve this problem, a neutrosophic judgment matrix can be decomposed into several neutrosophic sub-judgment matrixes. Each neutrosophic sub-judgment matrix has a lower indeterminacy than the original neutrosophic judgment matrix.

Keywords Neutrosophic collaborative intelligence · Neutrosophic collaborative decision-making · Neutrosophic analytic hierarchy process · Neutrosophic technique for order preference by similarity to an ideal solution

3.1 Neutrosophic Decision-Making

Neutrosophic logic can be embedded into various decision-making methods [1–12]. In particular, analytic hierarchy process (AHP)—the technique for order preference by similarity to an ideal solution (TOPSIS) combination—is among the most widely applied multi-criteria decision-making (MCDM) methods. Subsequently, several studies have applied neutrosophic AHP–TOPSIS.

T.-C.T. Chen, *Neutrosophic Collaborative Intelligence and Systems*,
SpringerBriefs in Applied Sciences and Technology,
https://doi.org/10.1007/978-3-032-34267-6_3

Nafei et al. [4] applied neutrosophic AHP–TOPSIS to green supplier selection using single-valued neutrosophic sets (SVNSs). They reviewed six score functions to convert the neutrosophic judgment matrix into a crisp matrix. After applying all six score functions, the arithmetic mean and TOPSIS were used to rank multiple green suppliers. The ranks of each green supplier across these lists were aggregated using a sophisticated neural network.

Razdan et al. [5] used neutrosophic AHP–TOPSIS to rank factors influencing supply chain risks. All the above researchers followed a similar procedure: replacing pairwise comparisons with score function values, approximating criterion weights using the arithmetic mean, and applying TOPSIS to evaluate and compare alternatives.

Ortega et al. [13] assessed water quality management in the Sinos River Basin from a socio-environmental perspective using the fuzzy c-means (FCM) and neutrosophic AHP–TOPSIS. Scenarios were first generated using the FCM and subsequently compared using neutrosophic AHP–TOPSIS. Each judgment was transformed using the score function proposed by Abdel-Basset et al. [12]. Thereafter, the arithmetic mean was applied to approximate criterion weights, and TOPSIS was used to evaluate the overall performance of the scenarios.

Junaid et al. [2] applied neutrosophic AHP–TOPSIS to supply chain risk assessment in the automotive industry in Pakistan. In their study, six supply chain risks were evaluated using three critical criteria: supply chain resilience, agility, and robustness. The score function used to convert the pairwise comparison results differed from that adopted by Ortega et al. [13]. Using TOPSIS, 17 alternatives were evaluated and compared based on these three criteria.

Nabeeh et al. [14] applied neutrosophic AHP–TOPSIS to personnel selection.

Reig Mullor et al. [15] evaluated the sustainability performance of a corporation, encompassing environmental, social, and governance dimensions, using neutrosophic AHP–TOPSIS. In that study, a new score function was proposed to convert each neutrosophic pairwise comparison result into a crisp value by incorporating the mean and standard deviation. Thereafter, the arithmetic mean and TOPSIS were applied to derive criterion weights and compare scenarios, respectively.

Andam et al. [16] applied neutrosophic AHP–TOPSIS to assess the performance of a healthcare service supply chain, considering eight factors including treatment brand, service-to-cost ratio, and technical comfort. Unlike most researchers that used the arithmetic mean to approximate criterion weights, they derived exact weights through eigenvalue analysis. To support this approach, various computational and optimization techniques were compared.

Atassi and Yang [17] applied the method to evaluate strategies for overcoming barriers to renewable energy development, using SVNSs.

Chiu and Chen [6] introduced a neutrosophic collaborative intelligence (NCI) approach designed to prioritize strategies for mitigating the challenges posed by cultural differences encountered by wafer foundries while establishing localized wafer fabs. This method emphasized a more rational aggregation of expert judgments, preserving indeterminacy in these evaluations until the final stage, thus enabling a more adaptable decision-making process. Within the NCI framework, each expert applied the calibrated neutrosophic geometric mean to compute neutrosophic weights for ranking the criteria related to mitigation measures. Subsequently, the neutrosophic technique for order of preference by similarity to the ideal solution was utilized to assess each measure's suitability. To consolidate the evaluations made by all experts, the approach introduced a neutrosophic weighted intersection (NWI) operator. The methodology also included a structured procedure for systematically prioritizing mitigation measures.

Chen and Chiu [18] propose an evolving neutrosophic fuzzy arithmetic mean-decomposition AHP–TOPSIS approach to prioritizing barriers to be addressed in semiconductor supply chain localization, in which fuzzy arithmetic mean was employed to decompose a highly indeterminate neutrosophic judgment matrix into multiple sub-judgment matrices with reduced indeterminacy. Subsequently, dominated sorting genetic algorithm (NSGA) was employed to simultaneously optimize consistency and indeterminacy. The neutrosophic weights of criteria were then derived from the sub-judgment matrix using calibrated neutrosophic geometric mean (cNGM). Finally, neutrosophic TOPSIS was used to evaluate the priority of each barrier.

3.2 Procedure for Neutrosophic Collaborative Decision-Making

Neutrosophic collaborative decision-making methods are the application of neutrosophic collaborative intelligence (NCI) in group decision-making. The procedure for neutrosophic collaborative decision-making comprises the following steps (see Fig. 3.1) [1]:

Step 1. All decision makers make their judgments (about the relative priorities of criteria): It is important to incorporate indeterminacy in each judgment. In some existing neutrosophic decision-making methods, judgments are given in linguistic terms that map to neutrosophic numbers.

Step 2. (Optional) Decision makers may deneutrosophicate their judgments: A common treatment in this field is to convert each neutrosophic judgment into a crisp one using the score function [6, 7].

Step 3. (Optional) Aggregate the judgments by all decision makers now or until Step 6.

Step 4. Each decision maker derives the neutrosophic or crisp weights of criteria.

Step 5. Evaluate the overall performance of each alternative.

Step 6. (Optional) Aggregate the evaluation results by all decision makers.

Step 7. Choose the top performing alternative.

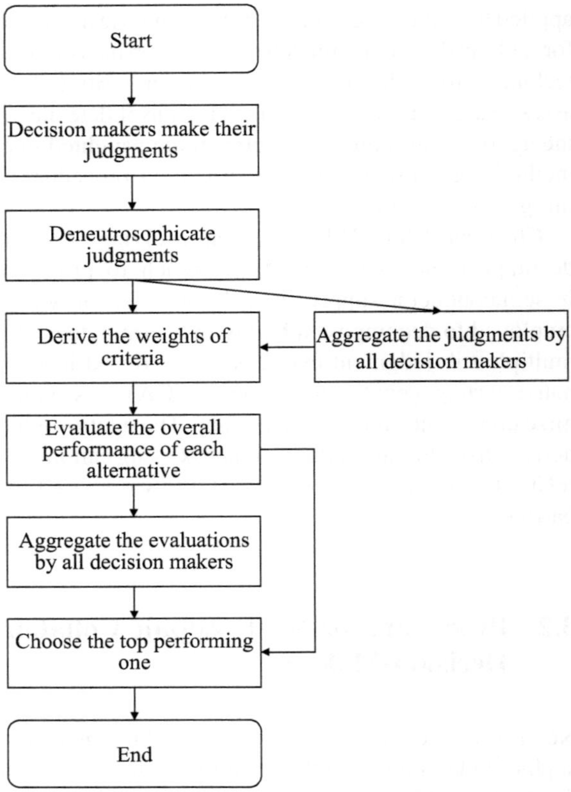

Fig. 3.1 Procedure for neutrosophic collaborative decision-making

3.3 Criterion Weight Derivation

3.3.1 Neutrosophic Judgments

In MCDM methods such as analytic hierarchy process (AHP) [8], best–worst method (BWM) [9], and measuring attractiveness by a categorical based evaluation technique (MACBETH) [10], decision makers make their judgments regarding the relative weights of criteria. Subsequently, the weights of criteria are derived from the judgment matrix comprising all judgments.

For example, a decision maker compares the relative weight of criterion i over criterion j with a linguistic term in Table 3.1, corresponding to a triangular neutrosophic number (TNN) $\tilde{a}_{ij}$; $i, j \in [1, n]$ [6]:

- In Table 3.1, some judgments are subject to higher indeterminacy, such as "weakly more important than," "as equal as or weakly more important than," and "as equal as," which imply the two criteria may be difficult to compare.

Table 3.1 TNNs for modeling linguistic terms

Linguistic Term	$\tilde{a}_{ij}$
As equal as	⟨(1, 1, 1), 0.5, 0.5, 0.5⟩
As equal as or weakly more important than	⟨(1, 2, 3), 0.4, 0.65, 0.6⟩
Weakly more important than	⟨(2, 3, 4), 0.3, 0.75, 0.7⟩
Weakly or strongly more important than	⟨(3, 4, 5), 0.6, 0.35, 0.4⟩
Strongly more important than	⟨(4, 5, 6), 0.8, 0.15, 0.2⟩
Strongly or very strongly more important than	⟨(5, 6, 7), 0.7, 0.25, 0.3⟩
Very strongly more important than	⟨(6, 7, 8), 0.9, 0.1, 0.1⟩
Very strongly or absolutely more important than	⟨(7, 8, 9), 0.85, 0.1, 0.15⟩
Absolutely more important than	⟨(9, 9, 9), 1, 0, 0⟩

- Another judgment with a similar property is "strongly more important than," which is in the middle of the scale (neither too high nor too low).
- Nevertheless, basically the higher the value of a TNN, the lower the indeterminacy it has.

$\tilde{\mathbf{A}} = [\tilde{a}_{ij}]$ is the neutrosophic judgment matrix or neutrosophic pairwise comparison matrix.

The arithmetic operations of TNNs with standard t-norm and standard t-conorm are provided in the following.

Theorem 3.1 (TNN Arithmetic Operations) [6]

- Addition:

$$\tilde{A}(+)\tilde{B} = \langle (A_1 + B_1,\ A_2 + B_2, A_3 + B_3);\ \min(\alpha_{\tilde{A}},\ \alpha_{\tilde{B}}), \max(\theta_{\tilde{A}},\ \theta_{\tilde{B}}), \max(\beta_{\tilde{A}},\ \beta_{\tilde{B}})\rangle$$

- Subtraction:

$$\tilde{A}(-)\tilde{B} = \langle (A_1 - B_3,\ A_2 - B_2, A_3 - B_1);\ \ \min(\alpha_{\tilde{A}},\ \alpha_{\tilde{B}}), \max(\theta_{\tilde{A}},\ \theta_{\tilde{B}}), \max(\beta_{\tilde{A}},\ \beta_{\tilde{B}})\rangle$$

- Inverse: $\frac{1}{\tilde{A}} \cong \left(\frac{1}{A_3}, \frac{1}{A_2}, \frac{1}{A_1}\right); \alpha_{\tilde{A}}, \theta_{\tilde{A}}, \beta_{\tilde{A}}$ if $A_1 \geq 0$
- Multiplication:

$$\tilde{A}(\times)\tilde{B} \cong (A_1B_1, A_2B_2, A_3B_3);\ \min(\alpha_{\tilde{A}}, \alpha_{\tilde{B}}), \max(\theta_{\tilde{A}}, \theta_{\tilde{B}}), \max(\beta_{\tilde{A}}, \beta_{\tilde{B}}) \quad \text{if } A_1, B_1 \geq 0$$

- Division:

$$\tilde{A}(/)\tilde{B} \cong (A_1/B_3, A_2/B_2, A_3/B_1);\ \min(\alpha_{\tilde{A}}, \alpha_{\tilde{B}}), \max(\theta_{\tilde{A}}, \theta_{\tilde{B}}), \max(\beta_{\tilde{A}}, \beta_{\tilde{B}}) \quad \text{if } B_1 \geq 0$$

Square root: $\sqrt{\tilde{A}} \cong (\sqrt{A_1}, \sqrt{A_2}, \sqrt{A_3}); \alpha_{\tilde{A}}, \theta_{\tilde{A}}, \beta_{\tilde{A}}$ if $A_1 \geq 0$.

3.3.2 Deneutrosophication of Neutrosophic Judgments

In some neutrosophic collaborative decision-making studies, judgments are deneutrosophicated before subsequent operations using a score function to reduce the difficulty of computing [11]. For example, a prevalent score function is

$$S(\tilde{a}_{ij}) = \frac{1}{8}(a_{ij1} + a_{ij2} + a_{ij3}) \cdot (2 + \alpha_{\tilde{a}_{ij}} - \theta_{\tilde{a}_{ij}} - \beta_{\tilde{a}_{ij}}) \tag{3.1}$$

Then, a crisp judgment matrix $\mathbf{A} = [S(\tilde{a}_{ij})]$ is formed. Other score functions are also applicable, such as [12]

$$S(\tilde{a}_{ij}) = \frac{1}{3}(a_{ij1} + a_{ij2} + a_{ij3}) \cdot (\alpha_{\tilde{a}_{ij}} - \theta_{\tilde{a}_{ij}} - \beta_{\tilde{a}_{ij}}) \tag{3.2}$$

The TNNs in Table 3.1 are deneutrosophicated according to Eq. (3.1), and the results are shown in Table 3.2. Their normalized values are also derived, so that the relative weight of "as equal as" is 1.

Example 3.1 The following neutrosophic judgment matrix.

$$\tilde{\mathbf{A}} = \begin{vmatrix} 1 & \langle(1, 2, 3), 0.4, 0.65, 0.6\rangle & \langle(3, 4, 5), 0.6, 0.35, 0.4\rangle \\ - & 1 & \langle(2, 3, 4), 0.3, 0.75, 0.7\rangle \\ - & - & 1 \end{vmatrix}$$

can be deneutrosophicated into the crisp judgment matrix below:

$$\mathbf{A} = \begin{vmatrix} 1 & 0.86 & 2.78 \\ - & 1 & 0.96 \\ - & - & 1 \end{vmatrix}$$

Table 3.2 Deneutrosophication results for the TNNs in Table 3.1

Linguistic term	Deneutrosophication result	Normalized value
As equal as	0.56	1.00
As equal as or weakly more important than	0.86	1.53
Weakly more important than	0.96	1.70
Weakly or strongly more important than	2.78	4.93
Strongly more important than	4.59	8.17
Strongly or very strongly more important than	4.84	8.60
Very strongly more important than	7.09	12.60
Very strongly or absolutely more important than	7.80	13.87
Absolutely more important than	10.13	18.00

Table 3.3 Score functions for TNNs used in past studies

Reference	Score Function
Abdel-Basset et al. [7], Ortega et al. [13], Razdan et al. [5]	$S(\tilde{A}) = \frac{1}{8}(A_1 + A_2 + A_3) \cdot (2 + \alpha_{\tilde{A}} - \theta_{\tilde{A}} - \beta_{\tilde{A}})$
Junaid et al. [2]	$S(\tilde{A}) = \frac{(A_1+A_2+A_3)}{3} + (\alpha_{\tilde{A}} - \theta_{\tilde{A}} - \beta_{\tilde{A}})$
Nabeeh et al. [14]	$S(\tilde{A}) = \frac{1}{9}(A_1 + A_2 + A_3) \cdot (\alpha_{\tilde{A}} + \theta_{\tilde{A}} + \beta_{\tilde{A}})$
Reig-Mullor et al. [15]	$S(\tilde{A}) = \frac{1}{18}(A_1 + 4A_2 + A_3) \cdot (2 + \alpha_{\tilde{A}} - \theta_{\tilde{A}} - \beta_{\tilde{A}}) + \frac{1}{\sqrt{72}}(A_3 - A_1) \cdot \left(\sqrt{\alpha_{\tilde{A}}} + \sqrt{1-\theta_{\tilde{A}}} + \sqrt{1-\beta_{\tilde{A}}}\right)$
Andam et al. [16]	$S(\tilde{A}) = \frac{1}{12}(A_1 + 2A_2 + A_3) \cdot (2 + \alpha_{\tilde{A}} - \theta_{\tilde{A}} - \beta_{\tilde{A}})$

which is easier to derive the (crisp) weights of criteria using eigen analysis.

Other score functions (see Table 3.3) are also applicable.

3.3.3 *Neutrosophic Geometric Mean*

Neutrosophic geometric mean (NGM), an extension of fuzzy geometric mean (FGM) [19], can be applied to derive the neutrosophic weights of criteria from a neutrosophic judgment matrix as

$$\tilde{w}_i(NGM) \cong \frac{\sqrt[n]{\prod_{j=1}^{n} \tilde{a}_{ij}}}{\sum_{k=1}^{n} \sqrt[n]{\prod_{j=1}^{n} \tilde{a}_{kj}}} = \frac{1}{1 + \frac{\sum_{k \neq i} \sqrt[n]{\prod_{j=1}^{n} \tilde{a}_{kj}}}{\sqrt[n]{\prod_{j=1}^{n} \tilde{a}_{ij}}}}; i = 1 \sim n \tag{3.3}$$

with the following membership functions (when standard t-norm and standard t-conorm are employed):

$$\alpha_{\tilde{w}_i} = \min_{j,k} \alpha_{\tilde{a}_{kj}} \tag{3.4}$$

$$\theta_{\tilde{w}_i} = \max_{j,k} \theta_{\tilde{a}_{kj}} \tag{3.5}$$

$$\beta_{\tilde{w}_i} = \max_{j,k} \beta_{\tilde{a}_{kj}} \tag{3.6}$$

Example 3.2 NGM is applied to derive the neutrosophic weights of criteria in Example 3.1 as follows:

$$\begin{aligned}
\tilde{w}_1(NGM) &= \frac{1}{1+\frac{\sqrt[3]{\frac{1}{(1,2,3)}(\times)1(\times)(2,3,4)}}{\sqrt[3]{1(\times)(1,2,3)(\times)(3,4,5)}}+\frac{\sqrt[3]{\frac{1}{(3,4,5)}(\times)\frac{1}{(2,3,4)}(\times)1}}{\sqrt[3]{1(\times)(1,2,3)(\times)(3,4,5)}}} \\
&\cong (0.403, 0.558, 0.665) \\
\tilde{w}_2(NGM) &\cong (0.225, 0.320, 0.467) \\
\tilde{w}_3(NGM) &\cong (0.083, 0.122, 0.192)
\end{aligned}$$

Their memberships are evaluated as

$$\begin{aligned}
\alpha_{\tilde{w}_1} &= \alpha_{\tilde{w}_2} = \alpha_{\tilde{w}_3} = 0.3 \\
\theta_{\tilde{w}_1} &= \theta_{\tilde{w}_2} = \theta_{\tilde{w}_3} = 0.75 \\
\beta_{\tilde{w}_1} &= \beta_{\tilde{w}_2} = \beta_{\tilde{w}_3} = 0.7
\end{aligned}$$

In sum,

$$\begin{aligned}
\tilde{w}_1(NGM) &= (0.403, 0.558, 0.665); 0.3, 0.75, 0.7 \\
\tilde{w}_2(NGM) &= (0.225, 0.320, 0.467); 0.3, 0.75, 0.7 \\
\tilde{w}_3(NGM) &= (0.083, 0.122, 0.192); 0.3, 0.75, 0.7
\end{aligned}$$

3.3.4 Neutrosophic Extent Analysis

Neutrosophic geometric mean (NEA), an extension of fuzzy extension analysis (FEA) [20], can be applied to derive the neutrosophic weights of criteria precisely from a neutrosophic judgment matrix as.

$$w_i(NEA) = \min_{j \neq i} V\left(\tilde{s}_i \geq \tilde{s}_j\right); i = 1 \sim n \tag{3.7}$$

where

$$V\left(\tilde{s}_i \geq \tilde{s}_j\right) = \begin{cases} 1 & if \quad s_{i2} \geq s_{j2} \\ 0 & if \quad s_{i3} \leq s_{j1} \\ \frac{s_{j1}-s_{i3}}{s_{i2}-s_{i3}-\left(s_{j2}-s_{j1}\right)} & otherwise \end{cases} \tag{3.8}$$

in which

$$\tilde{s}_i = \frac{\frac{1}{n}\sum_{j=1}^{n}\tilde{a}_{ij}}{\sum_{k=1}^{n}\left(\frac{1}{n}\sum_{j=1}^{n}\tilde{a}_{kj}\right)} \tag{3.9}$$

The weights of criteria derived using NEA are crisp values.

Example 3.3 NEA is applied to derive the neutrosophic weights of criteria in Example 3.1 as follows. First,

$$\tilde{s}_1 = (0.390, 0.535, 0.653)$$
$$\tilde{s}_2 = (0.235, 0.344, 0.482)$$
$$\tilde{s}_3 = (0.088, 0.121, 0.180)$$

Therefore,

$$V(\tilde{s}_1 \geq \tilde{s}_2) = 1$$
$$V(\tilde{s}_1 \geq \tilde{s}_3) = 1$$
$$V(\tilde{s}_2 \geq \tilde{s}_1) = 0.326$$
$$V(\tilde{s}_2 \geq \tilde{s}_3) = 1$$
$$V(\tilde{s}_3 \geq \tilde{s}_1) = 0$$
$$V(\tilde{s}_3 \geq \tilde{s}_2) = 0$$

Subsequently,

$$w_1(NEA) = 1$$
$$w_2(NEA) = 0.356$$
$$w_3(NEA) = 0$$

After normalization,

$$w_1(NEA) = 0.754$$
$$w_2(NEA) = 0.246$$
$$w_3(NEA) = 0$$

In contrast, product t-norm and product t-conorm can be employed for the same purpose [1]:

$$\alpha_{\tilde{w}_i} = \xi_n\left(\{\alpha_{\tilde{a}_{kj}}\}\right) \cdot \prod_j \alpha_{\tilde{a}_{ij}} \tag{3.10}$$

$$\theta_{\tilde{w}_i} = \theta_{\tilde{a}_i} + \prod_{k \neq i} \theta_{\tilde{a}_k} - \theta_{\tilde{a}_i} + \prod_{k \neq i} \theta_{\tilde{a}_k} \tag{3.11}$$

$$\beta_{\tilde{w}_i} = \beta_{\tilde{a}_i} + \prod_{k \neq i} \beta_{\tilde{a}_k} - \beta_{\tilde{a}_i} + \prod_{k \neq i} \beta_{\tilde{a}_k} \tag{3.12}$$

where

$$\xi_c(\{\alpha_{\tilde{a}_{kj}}\}) = \begin{cases} 0 & if \quad c = 0 \text{ or } i \\ \prod_j \alpha_{\tilde{a}_{cj}} + \xi_{c-1}(\{\alpha_{\tilde{a}_{kj}}\}) - \prod_j \alpha_{\tilde{a}_{cj}} \xi_{c-1}(\{\alpha_{\tilde{a}_{kj}}\}) & otherwise \end{cases} \tag{3.13}$$

$$\theta_{\tilde{a}_i} = \zeta_{\tilde{a}_{in}} \tag{3.14}$$

$$\zeta_{\tilde{a}_{ic}} = \begin{cases} 0 & if \quad c = 0 \\ \theta_{\tilde{a}_{ic}} + \zeta_{\tilde{a}_{i,c-1}} - \theta_{\tilde{a}_{ic}} \cdot \zeta_{\tilde{a}_{i,c-1}} & otherwise \end{cases} \tag{3.15}$$

$$\beta_{\tilde{a}_i} = \varsigma_{\tilde{a}_{in}} \tag{3.16}$$

$$\varsigma_{\tilde{a}_{ic}} = \begin{cases} 0 & if \quad c = 0 \\ \beta_{\tilde{a}_{ic}} + \varsigma_{\tilde{a}_{i,c-1}} - \beta_{\tilde{a}_{ic}} \cdot \varsigma_{\tilde{a}_{i,c-1}} & otherwise \end{cases} \tag{3.17}$$

Example 3.4 In the previous example, product t-norm and product t-conorm are employed instead. For example,

$$\alpha_{\sqrt[3]{\prod_{j=1}^{3} \tilde{a}_{1j}}} = 1 \cdot 0.4 \cdot 0.6 = 0.24$$

$$\alpha_{\sqrt[3]{\prod_{j=1}^{3} \tilde{a}_{2j}}} = 0.4 \cdot 1 \cdot 0.3 = 0.12$$

$$\alpha_{\sqrt[3]{\prod_{j=1}^{3} \tilde{a}_{3j}}} = 0.6 \cdot 0.3 \cdot 1 = 0.18$$

$$\alpha_{\sqrt[3]{\prod_{j=1}^{3} \tilde{a}_{2j}} + \sqrt[3]{\prod_{j=1}^{3} \tilde{a}_{3j}}} = 0.12 + 0.18 - 0.12 \cdot 0.18 = 0.28$$

$$\alpha_{\left(\sqrt[3]{\prod_{j=1}^{3} \tilde{a}_{2j}} + \sqrt[3]{\prod_{j=1}^{3} \tilde{a}_{3j}}\right) / \sqrt[3]{\prod_{j=1}^{3} \tilde{a}_{1j}}} = 0.28 \cdot 0.24 = 0.07$$

$$\alpha_{\tilde{w}_1} = \alpha_{1/\left(1 + \frac{\sqrt[3]{\prod_{j=1}^{3} \tilde{a}_{2j}} + \sqrt[3]{\prod_{j=1}^{3} \tilde{a}_{3j}}}{\sqrt[3]{\prod_{j=1}^{3} \tilde{a}_{1j}}}\right)} = 0.07$$

Similarly,

$$\alpha_{\tilde{w}_2} = \alpha_{1/\left(1 + \frac{\sqrt[3]{\prod_{j=1}^{3} \tilde{a}_{1j}} + \sqrt[3]{\prod_{j=1}^{3} \tilde{a}_{3j}}}{\sqrt[3]{\prod_{j=1}^{3} \tilde{a}_{2j}}}\right)} = 0.05$$

$$\alpha_{\tilde{w}_3} = \alpha_{1/\left(1+\frac{\sqrt[3]{\prod_{j=1}^{3}\tilde{a}_{1j}}+\sqrt[3]{\prod_{j=1}^{3}\tilde{a}_{2j}}}{\sqrt[3]{\prod_{j=1}^{3}\tilde{a}_{3j}}}\right)} = 0.06$$

Obviously, using standard t-norm and standard t-conorm, the memberships of different weights are equal, while using product t-norm and product t-conorm the memberships of different weights become unequal. In addition,

$$\theta_{\sqrt[3]{\prod_{j=1}^{3}\tilde{a}_{1j}}} = (0 + 0.65 - 0 \cdot 0.65) + 0.35 - (0 + 0.65 - 0 \cdot 0.65) \cdot 0.35 = 0.77$$

$$\theta_{\sqrt[3]{\prod_{j=1}^{3}\tilde{a}_{2j}}} = (0.65 + 0 - 0.65 \cdot 0) + 0.75 - (0.65 + 0 - 0.65 \cdot 0) \cdot 0.75 = 0.91$$

$$\theta_{\sqrt[3]{\prod_{j=1}^{3}\tilde{a}_{3j}}} = (0.35 + 0.75 - 0.35 \cdot 0.75) + 0 - (0.35 + 0.75 - 0.35 \cdot 0.75) \cdot 0 = 0.84$$

$$\theta_{\sqrt[3]{\prod_{j=1}^{3}\tilde{a}_{2j}}+\sqrt[3]{\prod_{j=1}^{3}\tilde{a}_{3j}}} = 0.91 \cdot 0.84 = 0.76$$

$$\theta_{\left(\sqrt[3]{\prod_{j=1}^{3}\tilde{a}_{2j}}+\sqrt[3]{\prod_{j=1}^{3}\tilde{a}_{3j}}\right)/\sqrt[3]{\prod_{j=1}^{3}\tilde{a}_{1j}}} = 0.76 + 0.77 - 0.76 \cdot 0.77 = 0.94$$

$$\theta_{\tilde{w}_1} = \theta_{1/\left(1+\frac{\sqrt[3]{\prod_{j=1}^{3}\tilde{a}_{2j}}+\sqrt[3]{\prod_{j=1}^{3}\tilde{a}_{3j}}}{\sqrt[3]{\prod_{j=1}^{3}\tilde{a}_{1j}}}\right)} = 0.94$$

Furthermore,

$$\beta_{\tilde{w}_1} = 0.93$$

Therefore, $\tilde{w}_1(NGM) = (0.403, 0.558, 0.665); 0.07, 0.94, 0.93$.

3.3.5 Deneutrosophication + Eigen Analysis

If all neutrosophic judgments have been deneutrosophicated, then an eigen analysis is applied to derive the crisp weights of criteria:

$$\det(\mathbf{A} - \lambda\mathbf{I}) = 0 \tag{3.18}$$

$$(\mathbf{A} - \lambda\mathbf{I})\mathbf{x} = 0 \tag{3.19}$$

where λ and $\mathbf{x}$ are the eigenvalue and eigenvector of $\mathbf{A}$, respectively.

Example 3.5 An eigen analysis is applied to derive the crisp weights of criteria in Example 3.1 based on the deneutrosophicated judgment matrix. The results are.

```
import numpy as np
from numpy.linalg import eig

a = np.array([[1, 0.86, 2.78],
              [1.16, 1, 0.96],
              [0.36, 1.04, 1]])
e_val,e_vec = eig(a)
print('E-value:', e_val)
print('E-vector', e_vec)

weights = e_vec[:,0]/sum(e_vec[:,0])
print('Weights', weights)
```

Fig. 3.2 Python code for implementing eigen analysis

$$w_1 = 0.432,\ w_2 = 0.335,\ w_3 = 0.233.$$

The required Python code is provided in Fig. 3.2.

A prevalent way is to approximate the weights of criteria using arithmetic mean (or row mean):

$$w_i = \frac{\sum_{j=1}^{n} S(\tilde{a}_{ij})}{\sum_{k=1}^{n}\sum_{j=1}^{n} S(\tilde{a}_{kj})} \tag{3.20}$$

Example 3.6 Arithmetic mean is applied to derive the crisp weights of criteria in Example 3.1 based on the deneutrosophicated judgment matrix. The results are.

$$w_1 = 0.457, w_2 = 0.307, w_3 = 0.236.$$

3.3.6 Neutrosophic Extent Analysis

In cNGM, the neutrosophic weight of criterion i is obtained using the following formulation:

$$\tilde{w}_i(\text{cNGM}) = \frac{w_i^c}{w_{i2}(\text{NGM})} \cdot \tilde{w}_i(\text{NGM});\ i = 1 \sim n, m = 1 \sim M \tag{3.21}$$

where $\{w_i^c(m)\}$ denotes the solution to the crisp eigen analysis problem. Therefore, we have

$$w_{i1} \cong w_i^c \cdot \frac{1 + \sum_{m \neq i} \frac{\sqrt[n]{\prod_{j=1}^{n} a_{mj2}}}{\sqrt[n]{\prod_{j=1}^{n} a_{ij2}}}}{1 + \sum_{m \neq i} \frac{\sqrt[n]{\prod_{j=1}^{n} a_{mj,3}}}{\sqrt[n]{\prod_{j=1}^{n} a_{ij1}}}};\ i = 1 \sim n, \tag{3.22}$$

$$w_{i2} \cong w_i^c \cdot \frac{1+\sum_{m\neq i}\frac{\sqrt[n]{\prod_{j=1}^{n} a_{mj2}}}{\sqrt[n]{\prod_{j=1}^{n} a_{ij2}}}}{1+\sum_{m\neq i}\frac{\sqrt[n]{\prod_{j=1}^{n} a_{mj2}}}{\sqrt[n]{\prod_{j=1}^{n} a_{ij2}}}}; i = 1 \sim n, \tag{3.23}$$

$$w_{i3} \cong w_i^c \cdot \frac{1+\sum_{m\neq i}\frac{\sqrt[n]{\prod_{j=1}^{n} a_{mj2}}}{\sqrt[n]{\prod_{j=1}^{n} a_{ij2}}}}{1+\sum_{m\neq i}\frac{\sqrt[n]{\prod_{j=1}^{n} a_{mj1}}}{\sqrt[n]{\prod_{j=1}^{n} a_{ij3}}}}; i = 1 \sim n \tag{3.24}$$

Example 3.7 A crisp eigen analysis yields $w_1^c = 0.558$ for the judgment matrix previously considered in Example 3.4. The corresponding neutrosophic weight is then calibrated using the cNGM method:

$$w_{11} \cong 0.558 \cdot \frac{0.390}{0.535} = 0.407$$
$$w_{12} \cong 0.558 \cdot \frac{0.535}{0.535} = 0.558$$
$$w_{11} \cong 0.558 \cdot \frac{0.653}{0.535} = 0.682$$

3.4 Alternative Overall Performance Evaluation

Neutrosophic TOPSIS (NTOPSIS) is a prevalent MCDM method for evaluating and comparing the overall performances of alternatives [1]. In NTOPSIS, $\tilde{\rho}_{qi}$ indicates the normalized performance of alternative q in optimizing criterion i. For the larger-the-better criteria,

$$\tilde{\rho}_{qi} \cong \frac{1}{\sqrt{1+\sum_{\phi\neq q}(\frac{\tilde{p}_{\phi i}}{\tilde{p}_{qi}})^2}} \tag{3.25}$$

when distributive normalization is used; $\tilde{p}_{qi}$ is the original performance. For the-smaller-the-better criteria,

$$\tilde{\rho}_{qi} \cong 1 - \frac{1}{\sqrt{1+\sum_{\phi\neq q}(\frac{\tilde{p}_{\phi i}}{\tilde{p}_{qi}})^2}} \tag{3.26}$$

Both $\tilde{p}_{qi}$ and $\tilde{\rho}_{qi}$ may not be subject to indeterminacy, i.e., indeterminacy is zero.

Example 3.8 In a MCDM problem, there are five alternatives to be evaluated and compared using NTOPSIS. The performances of the alternatives in optimizing five criteria are presented in Table 3.4, which are expressed in triangular fuzzy numbers

(TFNs) and therefore their maximum truth, minimum indeterminacy, and minimum falsity memberships are 1, 0, and 0, respectively. The performances are normalized using distributive normalization. For example,

$$\tilde{\rho}_{11} \cong \frac{1}{\sqrt{1+(\frac{(0,1,2)}{(0,1,2)})^2+(\frac{(4,5,5)}{(0,1,2)})^2+(\frac{(3,4,5)}{(0,1,2)})^2+(\frac{(3,4,5)}{(0,1,2)})^2}} = (0, 0.13, 0.324)$$

The results are summarized in Table 3.5.

Subsequently, the neutrosophic weights of criteria derived somehow (e.g., subjectively assigned, by NAHP) are multiplied to the normalized performances to derive the neutrosophic weighted scores as

$$\tilde{s}_{qi} \cong \tilde{w}_i(\times)\tilde{\rho}_{qi} \tag{3.27}$$

with the following memberships:

$$\alpha_{\tilde{s}_{qi}} = \min\left(\alpha_{\tilde{w}_i}, \alpha_{\tilde{\rho}_{qi}}\right) = \alpha_{\tilde{w}_i} \tag{3.28}$$

$$\theta_{\tilde{s}_{qi}} = \max\left(\theta_{\tilde{w}_i}, \theta_{\tilde{\rho}_{qi}}\right) = \theta_{\tilde{w}_i} \tag{3.29}$$

$$\beta_{\tilde{s}_{qi}} = \max\left(\beta_{\tilde{w}_i}, \beta_{\tilde{\rho}_{qi}}\right) = \beta_{\tilde{w}_i} \tag{3.30}$$

Two reference points, the ideal solution and the anti-ideal solution, are then specified as [21]

$$\tilde{\Lambda}_i^+ = \max_q \tilde{s}_{qi}; \max_q \alpha_{\tilde{s}_{qi}}, \min_q \theta_{\tilde{s}_{qi}}, \min_q \beta_{\tilde{s}_{qi}} \tag{3.31}$$

$$\left\langle \tilde{\Lambda}_i^+ = \max_q \tilde{s}_{qi}; \max_q \alpha_{\tilde{s}_{qi}}, \min_q \theta_{\tilde{s}_{qi}}, \min_q \beta_{\tilde{s}_{qi}} \right\rangle \tag{3.32}$$

The distances between each alternative and the two reference points are measured as [21]

$$\tilde{d}_q^+ \cong \sqrt{\sum_{i=1}^{n} (\tilde{\Lambda}_i^+(-)\tilde{s}_{qi})^2} \tag{3.33}$$

$$\tilde{d}_q^- \cong \sqrt{\sum_{i=1}^{n} (\tilde{s}_{qi}(-)\tilde{\Lambda}_i^-)^2} \tag{3.34}$$

where

Table 3.4 Performances of the alternatives in optimizing five criteria

q	$\tilde{p}_{q1}$	$\tilde{p}_{q2}$	$\tilde{p}_{q3}$	$\tilde{p}_{q4}$	$\tilde{p}_{q5}$
1	$\langle(0, 1, 2); 1, 0, 0\rangle$	$\langle(0, 0, 1); 1, 0, 0\rangle$	$\langle(1.5, 2.5, 3.5); 1, 0, 0\rangle$	$\langle(3, 4, 5); 1, 0, 0\rangle$	$\langle(3, 4, 5); 1, 0, 0\rangle$
2	$\langle(0, 1, 2); 1, 0, 0\rangle$	$\langle(0, 0, 1); 1, 0, 0\rangle$	$\langle(1.5, 2.5, 3.5); 1, 0, 0\rangle$	$\langle(3, 4, 5); 1, 0, 0\rangle$	$\langle(3, 4, 5); 1, 0, 0\rangle$
3	$\langle(4, 5, 5); 1, 0, 0\rangle$	$\langle(4, 5, 5); 1, 0, 0\rangle$	$\langle(3, 4, 5); 1, 0, 0\rangle$	$\langle(4, 5, 5); 1, 0, 0\rangle$	$\langle(0, 0, 1); 1, 0, 0\rangle$
4	$\langle(3, 4, 5); 1, 0, 0\rangle$	$\langle(0, 0, 1); 1, 0, 0\rangle$	$\langle(3, 4, 5); 1, 0, 0\rangle$	$\langle(1.5, 2.5, 3.5); 1, 0, 0\rangle$	$\langle(1.5, 2.5, 3.5); 1, 0, 0\rangle$
5	$\langle(3, 4, 5); 1, 0, 0\rangle$	$\langle(0, 1, 2); 1, 0, 0\rangle$	$\langle(3, 4, 5); 1, 0, 0\rangle$	$\langle(0, 0, 1); 1, 0, 0\rangle$	$\langle(1.5, 2.5, 3.5); 1, 0, 0\rangle$

Table 3.5 Normalized performances

q	$\tilde{\rho}_{q1}$	$\tilde{\rho}_{q2}$	$\tilde{\rho}_{q3}$	$\tilde{\rho}_{q4}$	$\tilde{\rho}_{q5}$
1	$\langle(0, 0.13, 0.324), 1, 0, 0\rangle$	$\langle(0.13, 0.324, 0), 1, 0, 0\rangle$	$\langle(0.324, 0, 0), 1, 0, 0\rangle$	$\langle(0, 0, 0.243), 1, 0, 0\rangle$	$\langle(0, 0.243, 0.159), 1, 0, 0\rangle$
2	$\langle(0, 0.13, 0.324), 1, 0, 0\rangle$	$\langle(0.13, 0.324, 0), 1, 0, 0\rangle$	$\langle(0.324, 0, 0), 1, 0, 0\rangle$	$\langle(0, 0, 0.243), 1, 0, 0\rangle$	$\langle(0, 0.243, 0.159), 1, 0, 0\rangle$
3	$\langle(0.465, 0.651, 0.762), 1, 0, 0\rangle$	$\langle(0.651, 0.762, 0.834), 1, 0, 0\rangle$	$\langle(0.762, 0.834, 0.981), 1, 0, 0\rangle$	$\langle(0.834, 0.981, 1), 1, 0, 0\rangle$	$\langle(0.981, 1, 0.328), 1, 0, 0\rangle$
4	$\langle(0.367, 0.521, 0.707), 1, 0, 0\rangle$	$\langle(0.521, 0.707, 0), 1, 0, 0\rangle$	$\langle(0.707, 0, 0), 1, 0, 0\rangle$	$\langle(0, 0, 0.243), 1, 0, 0\rangle$	$\langle(0, 0.243, 0.328), 1, 0, 0\rangle$
5	$\langle(0.367, 0.521, 0.707), 1, 0, 0\rangle$	$\langle(0.521, 0.707, 0), 1, 0, 0\rangle$	$\langle(0.707, 0, 0.196), 1, 0, 0\rangle$	$\langle(0, 0.196, 0.447), 1, 0, 0\rangle$	$\langle(0.196, 0.447, 0.328), 1, 0, 0\rangle$

$$\alpha_{\tilde{d}_q^+} = \min_i(\min(\alpha_{\tilde{s}_{qi}}, \alpha_{\tilde{\Lambda}_i^+})) \tag{3.35}$$

$$\theta_{\tilde{d}_q^+} = \max_i(\max(\theta_{\tilde{s}_{qi}}, \theta_{\tilde{\Lambda}_i^+})) \tag{3.36}$$

$$\beta_{\tilde{d}_q^+} = \max_i(\max(\beta_{\tilde{s}_{qi}}, \beta_{\tilde{\Lambda}_i^+})) \tag{3.37}$$

The neutrosophic overall performance of alternative q is evaluated as

$$\tilde{o}_q \cong \left\langle \frac{1}{\frac{\tilde{d}_q^+}{\tilde{d}_q^-} + 1}; \ \min\left(\alpha_{\tilde{d}_q^+}, \ \alpha_{\tilde{d}_q^-}\right), \ \max\left(\theta_{\tilde{d}_q^+}, \ \theta_{\tilde{d}_q^-}\right), \ \max\left(\beta_{\tilde{d}_q^+}, \ \beta_{\tilde{d}_q^-}\right)\right\rangle \tag{3.38}$$

$\tilde{o}_q$ is the larger the better.

Example 3.9 In the previous example, the neutrosophic weights of criteria are derived using NGM as.

$$\tilde{w}_1 = (0.214, 0.345, 0.459); 0.885, 0.113, 0.115$$
$$\tilde{w}_2 = (0.055, 0.088, 0.144); 0.885, 0.113, 0.115$$
$$\tilde{w}_3 = (0029, 0.043, 0.069); 0.885, 0.113, 0.115$$
$$\tilde{w}_4 = (0.257, 0.361, 0.503); 0.885, 0.113, 0.115$$
$$\tilde{w}_5 = (0.106, 0.163, 0.256); 0.885, 0.113, 0.115$$

The weighted scores of the alternative are then derived and summarized in Table 3.6.

The ideal solution and the anti-ideal solution are specified according to Eqs. (3.31) and (3.32), as shown in Table 3.7.

The distances between each alternative and the two reference points are measured. The measurements results are summarized in Table 3.8.

Finally, the neutrosophic overall performance of each alternative is evaluated. The results are presented in Table 3.9.

3.5 Aggregation of Multiple Decision Makers' Evaluations

3.5.1 Neutrosophic Weighted Average

In a neutrosophic collaborative decision-making problem, multiple decision makers evaluate the overall performance of each alternative from different viewpoints [22]. It is assumed that the evaluation by each decision maker contains the actual value (i.e., the actual overall performance of the alternative if it can be known). Then, various

Table 3.6 Weighted scores

q	$\tilde{s}_{q1}$	$\tilde{s}_{q2}$	$\tilde{s}_{q3}$	$\tilde{s}_{q4}$	$\tilde{s}_{q5}$
1	⟨({0, 0.045, 0.149} \right), 0.885, 0.113, 0.115}⟩	⟨{0.045, 0.149, 0} \right), 0.885, 0.113, 0.115}⟩	⟨({0.149, 0, 0} \right), 0.885, 0.113, 0.115}⟩	⟨({0, 0, 0.035} \right), 0.885, 0.113, 0.115}⟩	⟨({0, 0.035, 0.005} \right), 0.885, 0.113, 0.115}⟩
2	⟨({0, 0.045, 0.149} \right), 0.885, 0.113, 0.115}⟩	⟨({0.045, 0.149, 0} \right), 0.885, 0.113, 0.115}⟩	⟨({0.149, 0, 0} \right), 0.885, 0.113, 0.115}⟩	⟨({0, 0, 0.035} \right), 0.885, 0.113, 0.115}⟩	⟨({0, 0.035, 0.005} \right), 0.885, 0.113, 0.115}⟩
3	⟨({0.099, 0.224, 0.35} \right), 0.885, 0.113, 0.115}⟩	⟨({0.224, 0.35, 0.046} \right), 0.885, 0.113, 0.115}⟩	⟨({0.35, 0.046, 0.086} \right), 0.885, 0.113, 0.115}⟩	⟨({0.046, 0.086, 0.144} \right), 0.885, 0.113, 0.115}⟩	⟨({0.086, 0.144, 0.009} \right), 0.885, 0.113, 0.115}⟩
4	⟨({0.078, 0.18, 0.325} \right), 0.885, 0.113, 0.115}⟩	⟨({0.18, 0.325, 0} \right), 0.885, 0.113, 0.115}⟩	⟨({0.325, 0, 0} \right), 0.885, 0.113, 0.115}⟩	⟨({0, 0, 0.035} \right), 0.885, 0.113, 0.115} ⟩	⟨({0, 0.035, 0.009} \right), 0.885, 0.113, 0.115}⟩
5	⟨({0.078, 0.18, 0.325} \right), 0.885, 0.113, 0.115}⟩	⟨({0.18, 0.325, 0} \right), 0.885, 0.113, 0.115}⟩	⟨({0.325, 0, 0.017} \right), 0.885, 0.113, 0.115}⟩	⟨({0, 0.017, 0.065} \right), 0.885, 0.113, 0.115}⟩	⟨({0.017, 0.065, 0.009} \right), 0.885, 0.113, 0.115}⟩

Table 3.7 The ideal solution and the anti-ideal solution

i	1	2	3	4	5
$\tilde{\Lambda}_i^+$	⟨({0.099, 0.224, 0.35} \right), 0.885, 0.113, 0.115}⟩	⟨({0.224, 0.35, 0.046} \right), 0.885, 0.113, 0.115}⟩	⟨({0.35, 0.046, 0.086} \right), 0.885, 0.113, 0.115}⟩	⟨({0.046, 0.086, 0.144} \right), 0.885, 0.113, 0.115}⟩	⟨({0.086, 0.144, 0.009} \right), 0.885, 0.113, 0.115}⟩
$\tilde{\Lambda}_i^-$	⟨({0, 0, 0} \right), 0.885, 0.113, 0.115}⟩	⟨({0, 0, 0} \right), 0.885, 0.113, 0.115}⟩	⟨({0, 0, 0} \right), 0.885, 0.113, 0.115}⟩	⟨({0, 0, 0} \right), 0.885, 0.113, 0.115}⟩	⟨({0, 0, 0} \right), 0.885, 0.113, 0.115}⟩

Table 3.8 Distances between each alternative and the two reference points

q	$\tilde{d}_q^+$	$\tilde{d}_q^-$
1	$\langle (0.011, 0.204, 0.501), 0.885, 0.113, 0.115 \rangle$	$\langle (0.102, 0.215, 0.434), 0.885, 0.113, 0.115 \rangle$
2	$\langle (0.011, 0.205, 0.503), 0.885, 0.113, 0.115 \rangle$	$\langle (0.1, 0.21, 0.43), 0.885, 0.113, 0.115 \rangle$
3	$\langle (0, 0.105, 0.428), 0.885, 0.113, 0.115 \rangle$	$\langle (0.16, 0.331, 0.537), 0.885, 0.113, 0.115 \rangle$
4	$\langle (0.011, 0.156, 0.49), 0.885, 0.113, 0.115 \rangle$	$\langle (0.092, 0.221, 0.446), 0.885, 0.113, 0.115 \rangle$
5	$\langle (0.033, 0.246, 0.521), 0.885, 0.113, 0.115 \rangle$	$\langle (0.081, 0.191, 0.376), 0.885, 0.113, 0.115 \rangle$

Table 3.9 Neutrosophic overall performance of each alternative

q	$\tilde{o}_q$
1	$\langle (0.17, 0.513, 0.975), 0.885, 0.113, 0.115 \rangle$
2	$\langle (0.166, 0.506, 0.974), 0.885, 0.113, 0.115 \rangle$
3	$\langle (0.272, 0.759, 1), 0.885, 0.113, 0.115 \rangle$
4	$\langle (0.159, 0.586, 0.975), 0.885, 0.113, 0.115 \rangle$
5	$\langle (0.135, 0.437, 0.919), 0.885, 0.113, 0.115 \rangle$

mechanisms can be employed to aggregate their evaluations. The first aggregation mechanism is neutrosophic weighted average (NWA):

$$\tilde{o}_q = \frac{\sum_{m=1}^{M} \left(\omega_m \cdot \tilde{o}_q(m) \right)}{\sum_{m=1}^{M} \omega_m} \tag{3.39}$$

where ω_m is the weight of decision maker #m; $m = 1 \sim M$:

$$\alpha_{\tilde{o}_q} = \min_m \alpha_{\tilde{o}_q(m)} \tag{3.40}$$

$$\theta_{\tilde{o}_q} = \max_m \theta_{\tilde{o}_q(m)} \tag{3.41}$$

$$\beta_{\tilde{o}_q} = \max_m \beta_{\tilde{o}_q(m)} \tag{3.42}$$

Example 3.10 Three decision makers have applied NTOPSIS to compare five alternatives. Take the first alternative as an example. The neutrosophic overall performance of the alternative has been evaluated by these decision makers. The evaluation results are summarized in Table 3.10, and illustrated in Fig. 3.3. The weights of these decision makers are 0.35, 0.25, and 0.4, respectively. These neutrosophic overall performances are aggregated using NWA as.

$$\tilde{o}_q = 0.35 \cdot (0.170, 0.513, 0.975) + 0.25 \cdot (0.098, 0.367, 0.605)$$
$$+0.4 \cdot (0.204, 0.476, 0.685) = (0.166, 0.462, 0.767)$$

Table 3.10 Neutrosophic overall performances evaluated by three decision makers

m	$\tilde{o}_q(m)$
1	$\langle(0.170, 0.513, 0.975), 0.885, 0.113, 0.115\rangle$
2	$\langle(0.098, 0.367, 0.605), 0.779, 0.252, 0.221\rangle$
3	$\langle(0.204, 0.476, 0.685), 0.812, 0.197, 0.188\rangle$

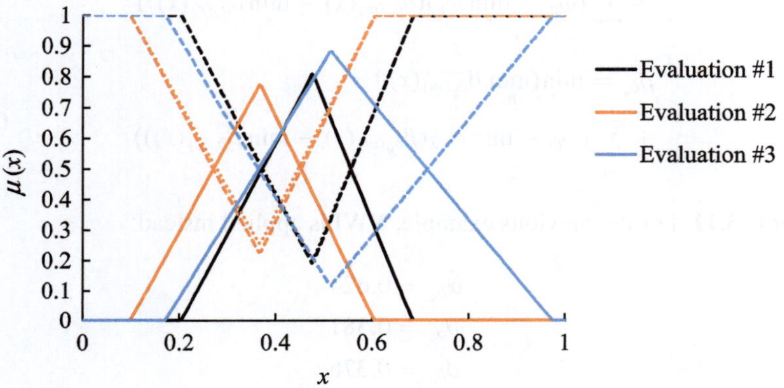

Fig. 3.3 Illustration of the evaluation results by three decision makers

$$\alpha_{\tilde{o}_q} = \min_m \alpha_{\tilde{o}_q(m)} = 0.779$$
$$\theta_{\tilde{o}_q} = \max_m \theta_{\tilde{o}_q(m)} = 0.252$$
$$\beta_{\tilde{o}_q} = \max_q \beta_{\tilde{o}_q(q)} = 0.221$$

Therefore, $\tilde{o}_q = (0.166, 0.462, 0.767); 0.779, 0.252, 0.221$.

3.5.2 Neutrosophic Weighted Intersection

Another aggregation mechanism is NWI. In NWI, the aggregation results of M neutrosophic evaluations, $\{\tilde{o}_q(m), m = 1 \sim M\}$, are aggregated as $\tilde{o}_q$, such that

$$\begin{aligned}\mu_{\tilde{o}_q}(x) &= \min_m \mu_{\tilde{o}_q(m)}(x) \\ &+ \sum_m (\omega_m - \min_r \omega_r)(\mu_{\tilde{o}_q(m)}(x) - \min_r \mu_{\tilde{o}_q(r)}(x))\end{aligned} \tag{3.43}$$

The membership functions of $\hat{y}_j$ are

$$\alpha_{\tilde{o}_q} = \max(\min_m \alpha_{\tilde{o}_q(m)}(x) + \sum_m (\omega_m - \min_r \omega_r)(\alpha_{\tilde{o}_q(m)}(x) - \min_r \alpha_{\tilde{o}_q(r)}(x))) \tag{3.44}$$

$$\theta_{\tilde{o}_q} = \min(\min_m \theta_{\tilde{o}_q(m)}(x) + \sum_m (\omega_m - \min_r \omega_r)(\theta_{\tilde{o}_q(m)}(x) - \min_r \theta_{\tilde{o}_q(r)}(x))) \tag{3.45}$$

$$\beta_{\tilde{o}_q} = \min(\min_m \beta_{\tilde{o}_q(m)}(x) + \sum_m (\omega_m - \min_r \omega_r)(\beta_{\tilde{o}_q(m)}(x) - \min_r \beta_{\tilde{o}_q(r)}(x))) \tag{3.46}$$

Example 3.11 For the previous example, NWI is applied instead:

$$\alpha_{\tilde{o}_q} = 0.624$$
$$\theta_{\tilde{o}_q} = 0.381$$
$$\beta_{\tilde{o}_q} = 0.376$$

The aggregation result is shown in Fig. 3.4.

Figure 3.5 compares the aggregation results of three neutrosophic evaluations using NWI, NWA, and NGM. In this example, the membership of 4.7 in the aggregation result using NWA or NGM is very high, while decision maker #2 does not think 4.7 is a possible value. The aggregation result is unreasonably biased toward other decision makers. NWI solves this problem by assigning a moderate membership to 4.7.

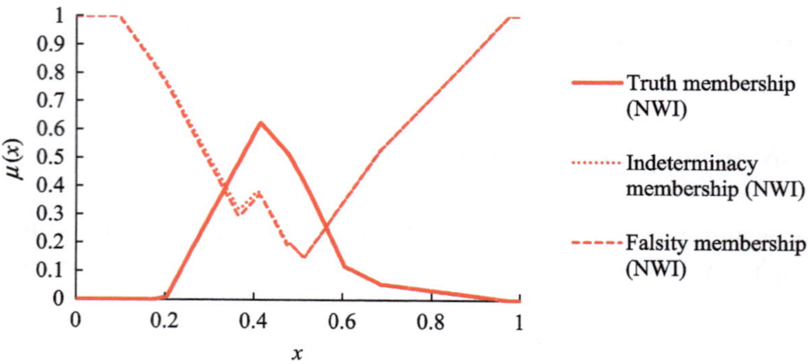

Fig. 3.4 Aggregation result using NWI [6]

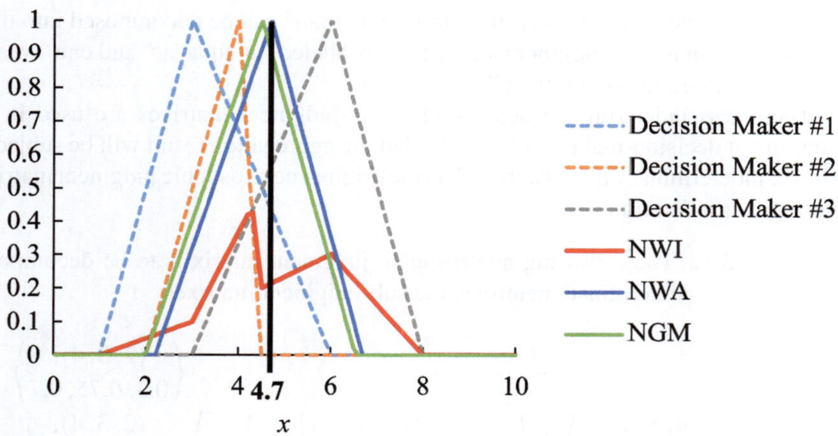

Fig. 3.5 Comparison of the aggregation results using NWI, NWA, and NGM

3.6 Reducing Indeterminacy After Aggregation Through Judgment Decomposition

Chen and Chiu [18] decomposed a neutrosophic judgment matrix into M neutrosophic sub-judgment matrixes using fuzzy arithmetic mean [23]. Each neutrosophic sub-judgment matrix had a lower indeterminacy than the original neutrosophic judgment matrix:

$$\tilde{\mathrm{A}} \cong \frac{\sum_{m=1}^{M} \tilde{\mathbf{A}}(m)}{M} \tag{3.47}$$

Specifically,

$$\tilde{a}_{ij} \cong \frac{\sum_{m=1}^{M} \tilde{a}_{ij}(m)}{M}, \text{ where } i, j \in [1, n] \tag{3.48}$$

such that

$$\theta_{\tilde{a}_{ij}} \geq \theta_{\tilde{a}_{ij}(m)};\ m = 1 \sim M \tag{3.49}$$

$$\widetilde{CR}\big(\tilde{\mathbf{A}}(m)\big) \leq 0.1;\ m = 1 \sim M \tag{3.50}$$

For example, the judgment "weakly more important than" can be decomposed into "as equal as or weakly more important than" and "weakly or strongly more important than." This decomposition reduces indeterminacy from 0.75 to the range 0.35–0.65.

A highly indeterminate judgment can be decomposed in multiple ways. Decomposing such a judgment into more than two less indeterminate sub-judgments is also feasible. For instance, “weakly more important than” can be decomposed into three less indeterminate sub-judgment matrices: two labeled “as equal as” and one “weakly or strongly more important than.”

These less indeterminate neutrosophic sub-judgment matrixes are used in the subsequent decision-making process, so that the aggregation result will be subject to a lower indeterminacy than that based on the original neutrosophic judgment matrixes with higher indeterminacy.

Example 3.12 The following neutrosophic judgment matrix is to be decomposed into two less indeterminate neutrosophic sub-judgment matrixes:

$$\tilde{\mathbf{A}} = \begin{bmatrix} 1 & - & - & - & \left\langle \begin{array}{c} (2,3,4), \\ 0.3, 0.75, 0.7 \end{array} \right\rangle \\ \left\langle \begin{array}{c} (4,5,6), \\ 0.8, 0.15, 0.2 \end{array} \right\rangle & 1 & \left\langle \begin{array}{c} (2,3,4), \\ 0.3, 0.75, 0.7 \end{array} \right\rangle & \left\langle \begin{array}{c} (1,1,1), \\ 0.5, 0.5, 0.5 \end{array} \right\rangle & \begin{array}{c} (2,3,4), \\ 0.3, 0.75, 0.7 \end{array} \\ \left\langle \begin{array}{c} (2,3,4), \\ 0.3, 0.75, 0.7 \end{array} \right\rangle & - & 1 & - & - \\ \left\langle \begin{array}{c} (2,3,4), \\ 0.3, 0.75, 0.7 \end{array} \right\rangle & - & \left\langle \begin{array}{c} (1,2,3), \\ 0.4, 0.65, 0.6 \end{array} \right\rangle & 1 & \left\langle \begin{array}{c} (2,3,4), \\ 0.3, 0.75, 0.7 \end{array} \right\rangle \\ - & - & \left\langle \begin{array}{c} (1,1,1), \\ 0.5, 0.5, 0.5 \end{array} \right\rangle & - & 1 \end{bmatrix}$$

Possible decomposition results are presented in Table 3.11.

Example 3.13 In Table 3.11, originally $\theta_{\tilde{w}_i} = \max_{i,j} \theta_{\tilde{a}_{ij}} = 0.75\, \forall i$. If $\tilde{\mathbf{A}}$ is decomposed into.

Table 3.11 Possible decomposition results

Judgment	TNN	Possible Decompositions
$\tilde{a}_{15}$	(2, 3, 4), 0.3, 0.75, 0.7	(1, 2, 3), 0.4, 0.65, 0.6 and (3, 4, 5), 0.6, 0.35, 0.4 (1, 1, 1), 0.5, 0.5, 0.5 and (4, 5, 6), 0.8, 0.15, 0.2
$\tilde{a}_{23}$	(2, 3, 4), 0.3, 0.75, 0.7	(1, 2, 3), 0.4, 0.65, 0.6 and (3, 4, 5), 0.6, 0.35, 0.4 (1, 1, 1), 0.5, 0.5, 0.5 and (4, 5, 6), 0.8, 0.15, 0.2
$\tilde{a}_{25}$	(2, 3, 4), 0.3, 0.75, 0.7	(1, 2, 3), 0.4, 0.65, 0.6 and (3, 4, 5), 0.6, 0.35, 0.4 (1, 1, 1), 0.5, 0.5, 0.5 and (4, 5, 6), 0.8, 0.15, 0.2
$\tilde{a}_{31}$	(2, 3, 4), 0.3, 0.75, 0.7	(1, 2, 3), 0.4, 0.65, 0.6 and (3, 4, 5), 0.6, 0.35, 0.4 (1, 1, 1), 0.5, 0.5, 0.5 and (4, 5, 6), 0.8, 0.15, 0.2
$\tilde{a}_{41}$	(2, 3, 4), 0.3, 0.75, 0.7	(1, 2, 3), 0.4, 0.65, 0.6 and (3, 4, 5), 0.6, 0.35, 0.4 (1, 1, 1), 0.5, 0.5, 0.5 and (4, 5, 6), 0.8, 0.15, 0.2
$\tilde{a}_{43}$	(1, 2, 3), 0.4, 0.65, 0.6	NA
$\tilde{a}_{45}$	(2, 3, 4), 0.3, 0.75, 0.7	(1, 2, 3), 0.4, 0.65, 0.6 and (3, 4, 5), 0.6, 0.35, 0.4 (1, 1, 1), 0.5, 0.5, 0.5 and (4, 5, 6), 0.8, 0.15, 0.2

$$\tilde{\mathbf{A}}(1) = \begin{bmatrix} 1 & - & - & - & \left\langle \begin{matrix}(1,2,3),\\ 0.4,0.65,0.6\end{matrix} \right\rangle \\ \left\langle \begin{matrix}(4,5,6),\\ 0.8,0.15,0.2\end{matrix} \right\rangle & 1 & \left\langle \begin{matrix}(1,2,3),\\ 0.4,0.65,0.6\end{matrix} \right\rangle & \left\langle \begin{matrix}(1,1,1),\\ 0.5,0.5,0.5\end{matrix} \right\rangle & \left\langle \begin{matrix}(1,2,3),\\ 0.4,0.65,0.6\end{matrix} \right\rangle \\ \left\langle \begin{matrix}(1,2,3),\\ 0.4,0.65,0.6\end{matrix} \right\rangle & - & 1 & - & - \\ \left\langle \begin{matrix}(1,2,3),\\ 0.4,0.65,0.6\end{matrix} \right\rangle & - & \left\langle \begin{matrix}(1,2,3),\\ 0.4,0.65,0.6\end{matrix} \right\rangle & 1 & \left\langle \begin{matrix}(1,2,3),\\ 0.4,0.65,0.6\end{matrix} \right\rangle \\ - & - & \left\langle \begin{matrix}(1,1,1),\\ 0.5,0.5,0.5\end{matrix} \right\rangle & - & 1 \end{bmatrix}$$

$$\tilde{\mathbf{A}}(2) = \begin{bmatrix} 1 & - & - & - & \left\langle \begin{matrix}(3,4,5),\\ 0.6,0.35,0.4\end{matrix} \right\rangle \\ \left\langle \begin{matrix}(4,5,6),\\ 0.8,0.15,0.2\end{matrix} \right\rangle & 1 & \left\langle \begin{matrix}(3,4,5),\\ 0.6,0.35,0.4\end{matrix} \right\rangle & \left\langle \begin{matrix}(1,1,1),\\ 0.5,0.5,0.5\end{matrix} \right\rangle & \left\langle \begin{matrix}(3,4,5),\\ 0.6,0.35,0.4\end{matrix} \right\rangle \\ \left\langle \begin{matrix}(3,4,5),\\ 0.6,0.35,0.4\end{matrix} \right\rangle & - & 1 & - & - \\ \left\langle \begin{matrix}(3,4,5),\\ 0.6,0.35,0.4\end{matrix} \right\rangle & - & \left\langle \begin{matrix}(1,2,3),\\ 0.4,0.65,0.6\end{matrix} \right\rangle & 1 & \left\langle \begin{matrix}(3,4,5),\\ 0.6,0.35,0.4\end{matrix} \right\rangle \\ - & - & \left\langle \begin{matrix}(1,1,1),\\ 0.5,0.5,0.5\end{matrix} \right\rangle & - & 1 \end{bmatrix}$$

then $\theta_{\tilde{w}_i}(1) = 0.65$ and $\theta_{\tilde{w}_i}(2) = 0.65$, both lower than the original indeterminacy.

Example 3.14 In Table 3.11, originally $\widetilde{CR}(\tilde{\mathbf{A}}) \cong 0.1$. After decomposition, $\widetilde{CR}(\tilde{\mathbf{A}}(1)) \cong 0.06$, while $\widetilde{CR}(\tilde{\mathbf{A}}(2)) \cong 0.15$. Therefore, another decomposition can be sought to guarantee both neutrosophic sub-judgment matrixes are consistent.

A neutrosophic judgment matrix can be decomposed in various ways (see Fig. 3.6). To identify the most suitable decomposition, the following bi-objective MINLP problem is solved [18]:

$$\text{Min } Z_1 = \sum_{m=1}^{M} CR_2(\tilde{\mathbf{A}}(m)) \tag{3.51}$$

$$\text{Min } Z_2 = \frac{1}{M} \sum_{m=1}^{M} IND(\tilde{\mathbf{A}}(m)) \tag{3.52}$$

s.t.

$$\tilde{\mathbf{A}} \cong \frac{\sum_{m=1}^{M} \tilde{\mathbf{A}}(m)}{M} \tag{3.53}$$

$$CR_2(\tilde{\mathbf{A}}(m)) \leq 0.1;\ m = 1 \sim M \tag{3.54}$$

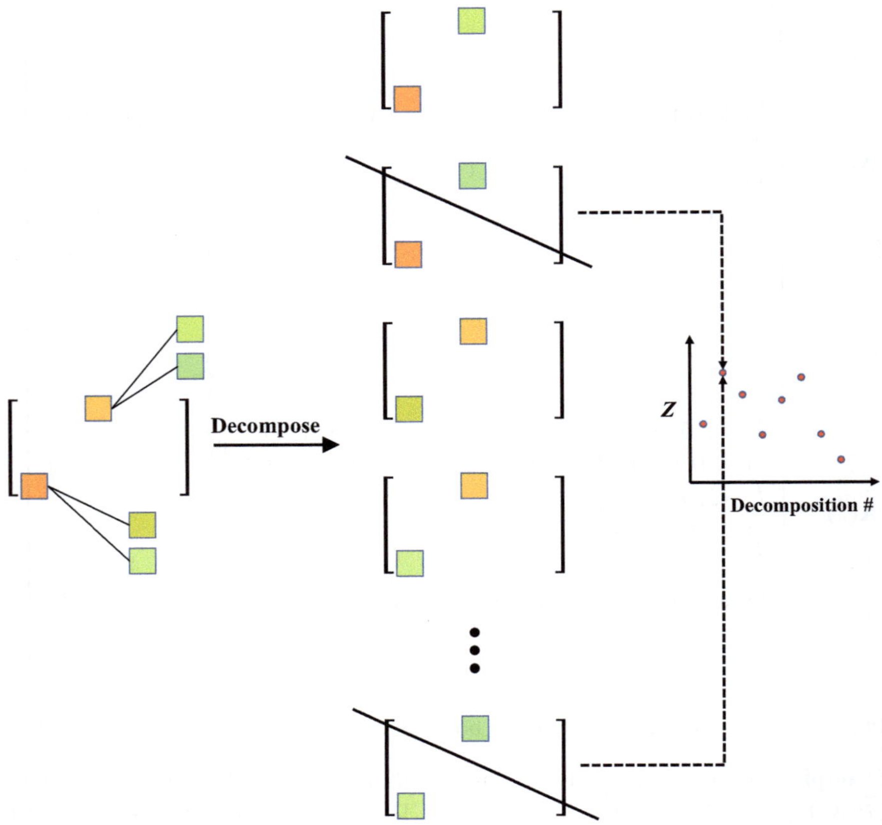

Fig. 3.6 Illustration of neutrosophic judgment matrix decomposition [18]

$$\theta_{\tilde{a}_{ij}(m)} \leq \theta_{\tilde{a}_{ij}}; i, j[1, n], m = 1 \sim M \tag{3.55}$$

The optimal solution yields M Pareto-optimal sub-judgment matrices.

To solve the bi-objective MINLP problem, Chen and Chiu [18] applied NSGA [24], as shown in Fig. 3.7.

The configuration of the NSGA involved the following parameters for population size, chromosome encoding, and crossover/mutation probabilities:

- Each chromosome was composed of C_2^n genes corresponding to the count of positive judgments.
- Gene values ranged from integers 1 to 9 (at most), each mapped to a specific linguistic term, as illustrated in Fig. 3.8: take $\tilde{a}_{14} = (2, 3, 4); 0.3, 0.75, 0.7$ as an example. According to Table 3.11, only four decompositions are possible. Therefore, the value of the corresponding gene is 0 to 4.

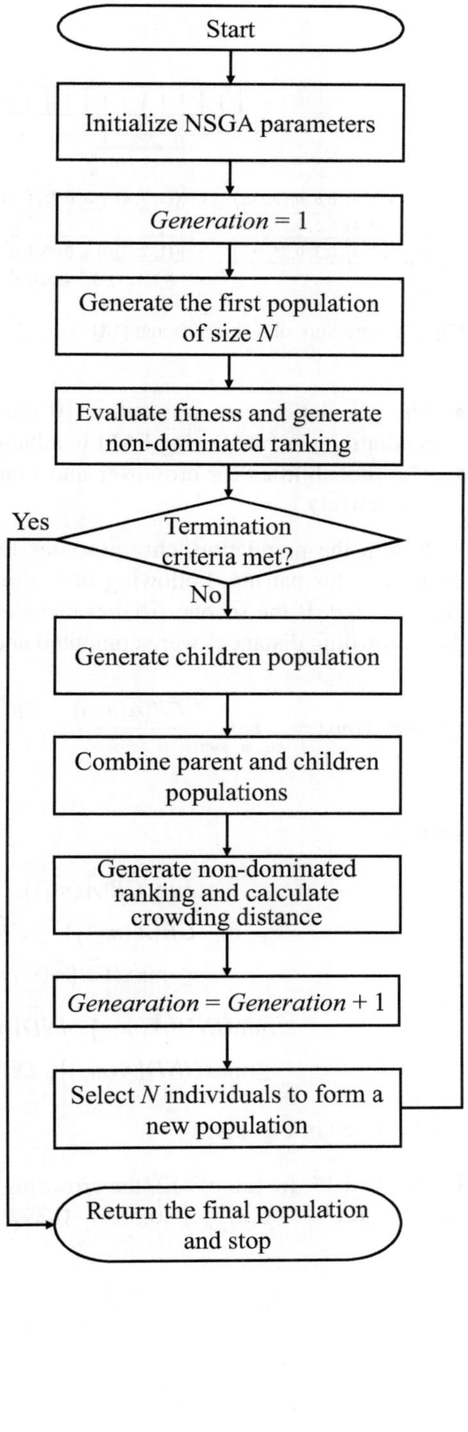

Fig. 3.7 NSGA for solving the bi-objective MINLP problem [18]

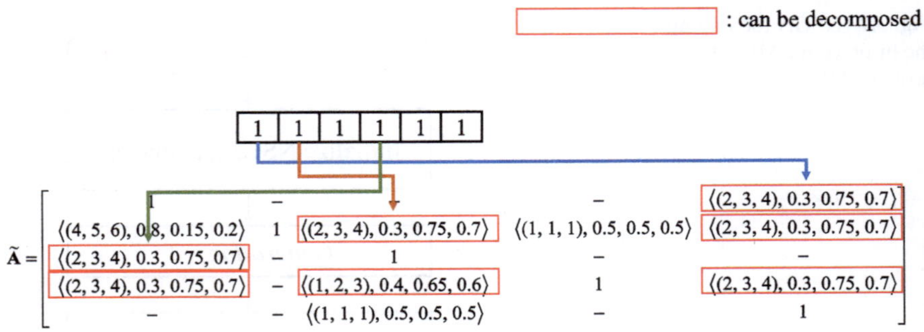

Fig. 3.8 Structure of a chromosome [18]

- The population size was set at 10 chromosomes, with the initial generation randomly generated using FAM in adherence to Constraints (3.50)–(3.51).
- The probabilities for crossover and mutation were established as 0.3 and 0.1, respectively.

Within the population, chromosomes located on the first Pareto frontier were prioritized for pairing. Following this, chromosomes on the second frontier were also selected. If the second frontier included a sufficient number of chromosomes, their crowding distances were computed and compared for further processing:

$$crowd\left(\tilde{\mathbf{A}}(m)\right) = \min_{m_1, m_2 \neq m} \frac{\left|\tilde{C}R\left(\tilde{\mathbf{A}}(m_1)\right) - \tilde{C}R\left(\tilde{\mathbf{A}}(m_2)\right)\right| + \left|IND\left(\tilde{\mathbf{A}}(m_1)\right) - IND\left(\tilde{\mathbf{A}}(m_2)\right)\right|}{2} \tag{3.56}$$

such that

$$\begin{aligned} &\min\left(\widetilde{CR}\left(\tilde{\mathbf{A}}(m_1)\right),\right. \\ &\left.\widetilde{CR}\left(\tilde{\mathbf{A}}(m_2)\right)\right) \leq \widetilde{CR}\left(\tilde{\mathbf{A}}(m)\right) \\ &\leq \max\left(\widetilde{CR}\left(\tilde{\mathbf{A}}(m_1)\right), \widetilde{CR}\left(\tilde{\mathbf{A}}(m_2)\right)\right) \end{aligned} \tag{3.57}$$

$$\begin{aligned} &\min\left(IND\left(\tilde{\mathbf{A}}(m_1)\right),\ IND\left(\tilde{\mathbf{A}}(m_2)\right)\right) \leq IND\left(\tilde{\mathbf{A}}(m)\right) \\ &\leq \max\left(IND\left(\tilde{\mathbf{A}}(m_1)\right),\ IND\left(\tilde{\mathbf{A}}(m_2)\right)\right) \end{aligned} \tag{3.58}$$

as illustrated in Fig. 3.9.

Example 3.15 In Table 3.12, the crowding distance of the Pareto-optimal solution $\tilde{\mathbf{A}}(3)$ is $(|0.092 - 0.079| + |0.565 - 0.392|)/2 = 0.093$.

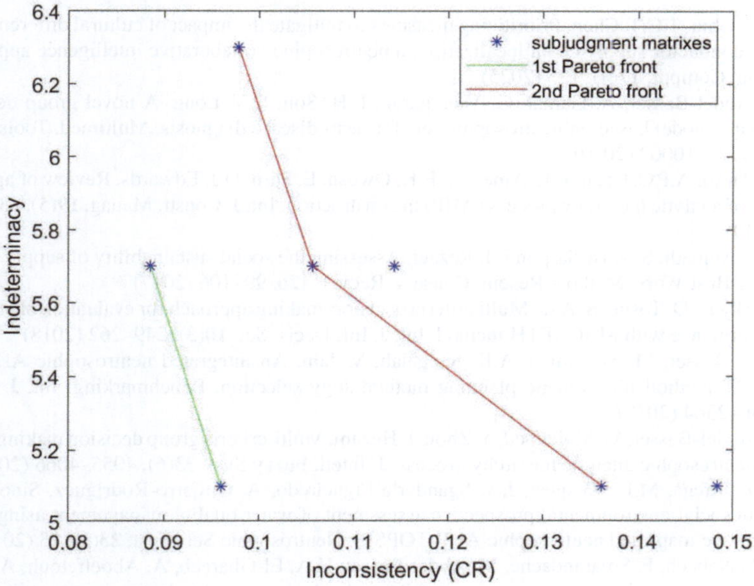

Fig. 3.9 First and second Pareto frontiers [18]

Table 3.12 Example for crowding distance calculation

m	$\widetilde{CR}(\tilde{\mathbf{A}}(m))$	$IND(\tilde{\mathbf{A}}(m))$
1	0.079	0.565
2	0.092	0.392
3	0.085	0.477

References

1. S.K. Patil, R. Kant, A fuzzy AHP-TOPSIS framework for ranking the solutions of Knowledge Management adoption in Supply Chain to overcome its barriers. Expert Syst. Appl. **41**(2), 679–693 (2014)
2. M. Junaid, Y. Xue, M.W. Syed, J.Z. Li, M. Ziaullah, A neutrosophic ahp and topsis framework for supply chain risk assessment in automotive industry of Pakistan. Sustainability **12**(1), 154 (2019)
3. M. Ortiz-Barrios, C. Miranda-De la Hoz, P. López-Meza, A. Petrillo, F. De Felice, A case of food supply chain management with AHP, DEMATEL, and TOPSIS. J. Multi-Criteria Decis. Anal. **27**(1–2), 104–128 (2020)
4. A. Nafei, S.P. Azizi, S.A. Edalatpanah, C.Y. Huang, Smart TOPSIS: a neural network-driven TOPSIS with neutrosophic triplets for green supplier selection in sustainable manufacturing. Expert Syst. Appl. **255**, 124744 (2024)
5. M.R. Razdan, S. Aghasi, S.M.R. Davoodi, Ranking factors affecting supply chain risk with a combined approach of neutrosophic analytical hierarchy process and TOPSIS. J. Appl. Res. Ind. Eng. **11**(3), 423–435 (2024)

6. M.C. Chiu, T.C.T. Chen, Prioritizing measures to mitigate the impact of cultural differences on semiconductor supply chain localization: a neutrosophic collaborative intelligence approach. Cogn. Comput. **17**(4), 125 (2025)
7. M. Abdel-Basset, A. Gamal, G. Manogaran, L.H. Son, H.V. Long, A novel group decision making model based on neutrosophic sets for heart disease diagnosis. Multimed. Tools Appl. **79**, 9977–10002 (2020)
8. A. Darko, A.P.C. Chan, E.E. Ameyaw, E.K. Owusu, E. Pärn, D.J. Edwards, Review of application of analytic hierarchy process (AHP) in construction. Int. J. Constr. Manag. **19**(5), 436–452 (2019)
9. H.B. Ahmadi, S. Kusi-Sarpong, J. Rezaei, Assessing the social sustainability of supply chains using Best Worst Method. Resour. Conserv. Recycl. **126**, 99–106 (2017)
10. G. Akyüz, Ö. Tosun, S. Aka, Multi criteria decision-making approach for evaluation of supplier performance with MACBETH method. Int. J. Inf. Decis. Sci. **10**(3), 249–262 (2018)
11. M.A. Basset, M. Mohamed, A.K. Sangaiah, V. Jain, An integrated neutrosophic AHP and SWOT method for strategic planning methodology selection. Benchmarking: Int. J. **25**(7), 2546–2564 (2018)
12. M. Abdel-Basset, M. Mohamed, Y. Zhou, I. Hezam, Multi-criteria group decision making based on neutrosophic analytic hierarchy process. J. Intell. Fuzzy Syst. **33**(6), 4055–4066 (2017)
13. R.G. Ortega, M.L. Vázquez, J.A. Sganderla Figueiredo, A. Guijarro-Rodriguez, Sinos river basin social-environmental prospective assessment of water quality management using fuzzy cognitive maps and neutrosophic AHP-TOPSIS. Neutrosophic Sets Syst. **23**(1), 13 (2018)
14. N.A. Nabeeh, F. Smarandache, M. Abdel-Basset, H.A. El-Ghareeb, A. Aboelfetouh, An integrated neutrosophic-topsis approach and its application to personnel selection: a new trend in brain processing and analysis. IEEE Access **7**, 29734–29744 (2019)
15. J. Reig-Mullor, A. Garcia-Bernabeu, D. Pla-Santamaria, M. Vercher-Ferrandiz, Evaluating ESG corporate performance using a new neutrosophic AHP-TOPSIS based approach. Technol. Econ. Dev. Econ. **28**(5), 1242–1266 (2022)
16. F. Andam, E. Asgharizadeh, M. Taghizadeh-Yazdi, Designing a model for healthcare services supply chain performance evaluation using neutrosophic multiple attribute decision-making technique. Int. J. Nonlinear Anal. Appl. **15**(9), 307–318 (2024)
17. R. Atassi, K. Yang, An integrated neutrosophic AHP and TOPSIS methods for assessment renewable energy barriers for sustainable development. Int. J. Neutrosophic Sci. **18**(2), 158–173 (2022)
18. T.C.T. Chen, M.C. Chiu, Prioritising barriers to be addressed in semiconductor supply chain localisation using an evolving neutrosophic multi-criteria decision-making approach. J. Oper. Res. Soc. 1–20 (2026)
19. T. Chen, Assessing factors critical to smart technology applications in mobile health care - the FGM-FAHP approach. Health Policy Technol. **9**, 194–203 (2020)
20. S. Ahmed, S. Ahmed, M.R.H. Shumon, E. Falatoonitoosi, M.A. Quader, A comparative decision-making model for sustainable end-of-life vehicle management alternative selection using AHP and extent analysis method on fuzzy AHP. Int J Sust Dev World **23**(1), 83–97 (2016)
21. T. Chen, H.-C. Wu, K.-W. Hsu, A FAHP-EFGM-FTOPSIS approach for suitable hotel recommendation amid the COVID-19 pandemic. Digital Health **8**, 20552076221084456 (2022)
22. T.C.T. Chen, H. Katsuhiro, Fuzzy collaborative intelligence. in *Explainable Fuzzy Collaborative Intelligence: Methodology, System Architecture, and Applications in Engineering* (2026), pp. 1–28
23. Y.C. Wang, T.C.T. Chen, A partial-consensus posterior-aggregation FAHP method—supplier selection problem as an example. Mathematics **7**(2), 179 (2019)
24. H.C. Fu, P. Liu, A multi-objective optimization model based on non-dominated sorting genetic algorithm. Int. J. Simul. Model. **18**(3), 510–520 (2019)

Chapter 4
Neutrosophic Collaborative Classification

Abstract This chapter begins by introducing the concept of neutrosophic classification and reviewing existing neutrosophic classification methods. Most existing neutrosophic classification methods convert input attributes into neutrosophic sets before classification. However, this treatment can be problematic because these input attributes may not possess uncertainty. To address this shortcoming, this chapter introduces three neutrosophic classification methods that use the crisp values of example attributes as inputs: neutrosophic c-means clustering, neutrosophic classification trees, and neutrosophic deep neural networks for classification. After describing the implementation process of such neutrosophic classification methods, this chapter introduces the three neutrosophic classification methods one by one. Subsequently, this chapter proposes a neutrosophic collaborative classification method that aggregates the classification results of multiple neutrosophic classification methods.

Keywords Neutrosophic collaborative intelligence · Neutrosophic collaborative classification · Neutrosophic c-means · Neutrosophic classification tree · Neutrosophic deep neural network

4.1 Neutrosophic Classification

A neutrosophic classification method can be the extension of a crisp classification method by deriving the three memberships or the extension of a fuzzy classification method by deriving additionally the indeterminacy membership, which then serves as the basis of collaboration, as shown in Fig. 4.1. For example, by measuring and comparing the distances from an item to various categories, the truth and falsity memberships of the item can be determined. In addition, the items of a category do not necessarily have the same actual classification results. Such deviations can be considered in determining the indeterminacy membership.

In the literature, there have been a few neutrosophic classification methods and their applications. For example, for leaf disease detection and classification, Dhingra

T.-C.T. Chen, *Neutrosophic Collaborative Intelligence and Systems*,
SpringerBriefs in Applied Sciences and Technology,
https://doi.org/10.1007/978-3-032-34267-6_4

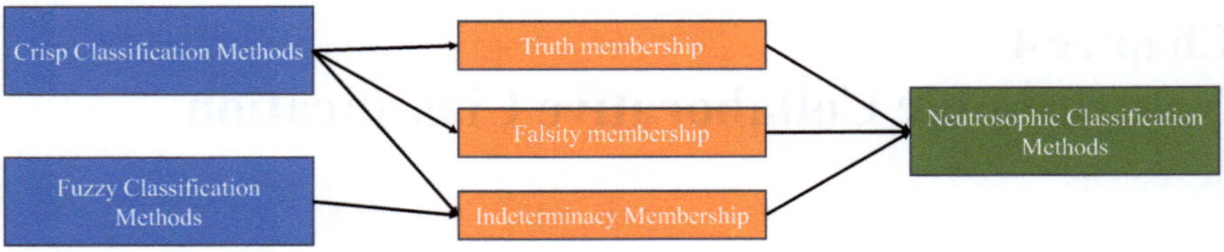

Fig. 4.1 From crisp and fuzzy classification methods to neutrosophic classification methods

et al. [1] proposed a segmentation technique based on neutrosophic logic to evaluate regions of interest, in which the segmented neutrosophic images were divided into three categories: true regions, false regions, and indeterminate regions. Based on the segmented regions, they evaluated the use of feature subsets such as texture, color, histogram, and disease sequence regions to identify the disease or health status of leaves. In addition, they used nine classifiers to monitor and verify the discriminative power of combined features, among which random forest classifier outperformed the other classifiers.

Ansari et al. [2] extended fuzzy classifiers to neutrosophic classifiers. To this end, they developed three fuzzy inference systems (FISs): one for modeling the truth part, one for modeling the indeterminacy part, and a third for modeling the falsity part, of each input feature. The indeterminacy membership was derived as follows (see Fig. 4.2):

$$\mu_{\tilde{x}_k}(x)(\text{determinacy}) = \max\left(\mu_{\tilde{x}_k}(x), \mu_{\tilde{x}_{k\pm1}}(x)\right) - \min\left(\mu_{\tilde{x}_k}(x), \mu_{\tilde{x}_k \pm 1}(x)\right) \quad (4.1)$$

$$\mu_{\tilde{x}_k}(x)(\text{indeterminacy}) = 1 - \mu_{\tilde{x}_k}(\text{determinacy}) \quad (4.2)$$

For example, the truth, indeterminacy, and falsity memberships of $x = 5$ in $\tilde{x}_1$ are 1, 0, and 0, respectively. In contrast, the memberships of $x = 7.5$ in $\tilde{x}_1$ are 0.5, 1, and 0.5, respectively. Although the operations of these components were independent of each other, there was a correlation between the membership functions of the neutrosophic intelligence truth, indeterminacy, and falsity components, which enabled the capture of the truth, indeterminacy, and falsity values of the inputs and output.

4.2 Existing Neutrosophic Forecasting Methods

Kandasamy and Smarandache [3] proposed a neutrosophic classification method for personality classification, in which people were requested to answer 31 questions. Their answers were mapped to triple refined indeterminate neutrosophic sets. Then, the distance between each pair of them was measured based on the proposed arithmetic operations for triple refined indeterminate neutrosophic sets. The measurement results were based on in classifying these people.

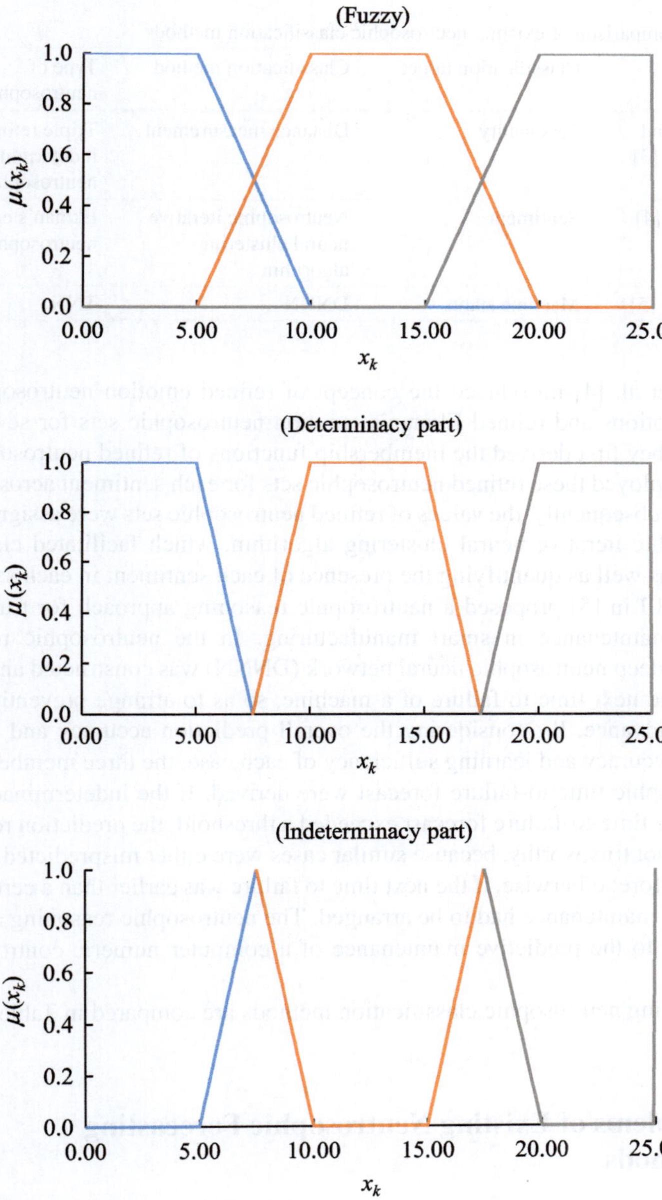

Fig. 4.2 Deriving the indeterminacy membership

Table. 4.1 Comparison of existing neutrosophic classification methods

Reference	Classification target	Classification method	Type of neutrosophic set
Kandasamy and Smarandache [3]	Personality	Distance measurement	Triple refined indeterminate neutrosophic set
Sharma et al. [4]	Sentiment	Neutrosophic iterative neural clustering algorithm	Ekman's emotion neutrosophic set
Chen and Lin [5]	Machine status	DNNN	TNN

Sharma et al. [4] introduced the concept of refined emotion neutrosophic sets for four emotions and refined Ekman's emotion neutrosophic sets for seven other emotions. They first derived the membership functions of refined neutrosophic sets and then employed these refined neutrosophic sets for each sentiment across a given taxonomy. Subsequently, the values of refined neutrosophic sets were assigned using a neutrosophic iterative neural clustering algorithm, which facilitated classifying sentiments as well as quantifying the presence of each sentiment in each sample.

Chen and Lin [5] proposed a neutrosophic reasoning approach for trustworthy predictive maintenance in smart manufacturing. In the neutrosophic reasoning approach, a deep neutrosophic neural network (DNNN) was constructed and trained to predict the next time to failure of a machine, so as to arrange preventive maintenance in advance. By considering the overall prediction accuracy and the local prediction accuracy and learning sufficiency of each case, the three memberships of the neutrosophic time-to-failure forecast were derived. If the indeterminacy of the neutrosophic time-to-failure forecast exceeded a threshold, the prediction result was considered not trustworthy, because similar cases were either mispredicted or rarely predicted before; otherwise, if the next time to failure was earlier than a certain date, a preventive maintenance had to be arranged. The neutrosophic reasoning approach was applied to the predictive maintenance of a computer numeric control (CNC) tool.

The existing neutrosophic classification methods are compared in Table 4.1.

4.3 Problems of Existing Neutrosophic Forecasting Methods

In comparison to the use of neutrosophic collaborative intelligence in fields like neutrosophic collaborative management, evaluation, and decision-making, there is a noticeably smaller body of references on neutrosophic collaborative classification. This highlights a clear lack of sufficient materials available for consultation in this specific field.

Most existing neutrosophic classification methods transform input attributes into neutrosophic sets before classification. However, such a treatment may be questionable since these input attributes may not be subject to uncertainty. In contrast, the classification mechanism itself is inherently uncertain. A neutrosophic classification method, which accepts the crisp values of example attributes as inputs, is detailed below.

4.4 Neutrosophic Classification Methods with Crisp Inputs

4.4.1 Procedure

The procedure for a neutrosophic classification method with crisp inputs comprises the following steps:

Step 1. Build and train a classification model to classify an item into one of several categories.

Step 2. Apply the trained classification model to test data.

Step 3. Derive the truth and falsity memberships of the neutrosophic classification result for each item in test data, based on the classification performance achieved by the classification model for the training data.

Step 4. Measure the local learning sufficiency (i.e., neighborhood crowding degree) of each item in test data.

Step 5. Evaluate the local classification performance for each item in test data.

Step 6. Derive the indeterminacy membership of the neutrosophic classification result for each item in test data, based on the local learning sufficiency and local classification performance.

Step 7. If the indeterminacy membership is less than a threshold specified in advance, the classification result is trustworthy; otherwise, the classification result is not trustworthy, and go to Step 9.

Step 8. Conduct SHAP analysis for generating a local explanation if an item is correctly classified; apply LIME instead if otherwise: Both are based on the premise that the classification result is trustworthy.

Step 9. End.

A flowchart is provided in Fig. 4.3 to illustrate the procedure of the neutrosophic classification method.

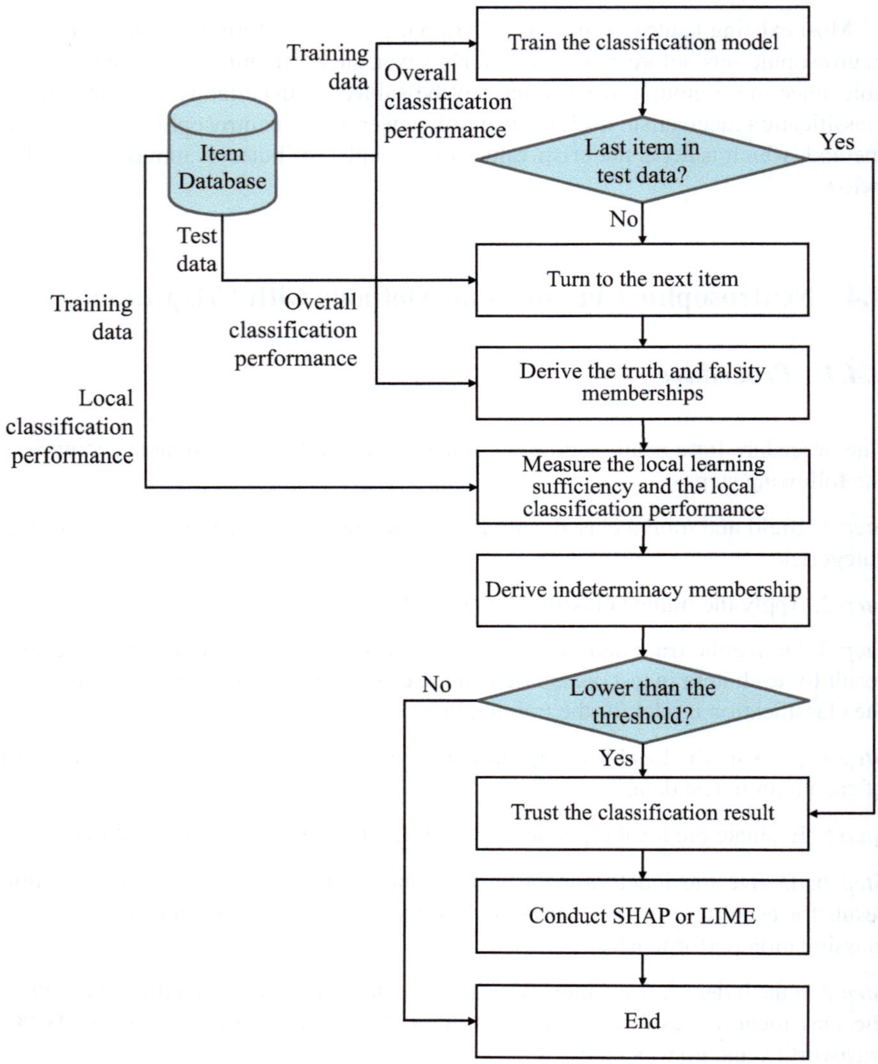

Fig. 4.3 Flowchart showing the procedure for implementing a neutrosophic classification method

4.4.2 *Neutrosophic c-Means*

Neutrosophic c-means is an extension from fuzzy c-means, a prevalent method for unsupervised clustering [6], which comprises the following steps:

Step 1. Establish an initial classification result.

Step 2. (Iterations) Obtain the centroid of each category as

$$\bar{\mathbf{x}}_{(c)} = \{\bar{x}_{(c)k}\} \quad (4.3)$$

where

$$\bar{x}_{(c)k} = \frac{\sum_{j=1}^{n} \mu_{j(c)}^{m} x_{jk}}{\sum_{j=1}^{n} \mu_{j(c)}^{m}} \quad (4.4)$$

$$\mu_{j(c)} = \frac{1}{\sum_{l=1}^{c} \left(\frac{e_{j(c)}}{e_{j(l)}}\right)^{\frac{2}{m-1}}} \quad (4.5)$$

$$e_{j(c)} = \sqrt{\sum_{k=1}^{K} (x_{jk} - \bar{x}_{(c)k})^2} \quad (4.6)$$

Step 3. Re-measure the distance of each example to the centroid of every category, and then recalculate the corresponding membership.

Step 4. Stop if the following condition is satisfied. Otherwise, return to Step 2:

$$\max_{c} \max_{j} |\mu_{j(c)}(t) - \mu_{j(c)}(t-1)| < d \quad (4.7)$$

where $\mu_{j(c)}(t)$ is the membership of example j belonging to category c after the t-th iteration; $d \in \mathbf{R}^{+}$ is the threshold of membership convergence.

Step 5. Derive the three memberships: Example j is classified into category $\arg\max_{c} \mu_{j(c)}$. The truth and falsity memberships of the classification result are set to $\max_{c} \mu_{j(c)}$ and $1 - \max_{c} \mu_{j(c)}$, respectively. If the example is absolutely classified into category c, i.e., $\mu_{j(c)} = 1$ while $\mu_{j(l)} = 0\ \forall l \neq c$, then the indeterminacy membership $\theta_j = 0$. In contrast, if the example cannot be classified into any category, $\mu_{j(c)} = \frac{1}{C}$ $\forall c$, then $\theta_j = 1$.

Theorem 1: A linear function for deriving the indeterminacy membership is.

$$I_j = \frac{-C}{C-1} \max_{c} \mu_{j(c)} + \frac{C}{C-1} \quad (4.8)$$

Proof: This is a linear regression problem [7] in which the fitted regression line passes two points (1, 0) and $(\frac{1}{C}, 1)$.

Example 4.1. The data of two hundred lots have been collected for cycle time prediction, as presented in Table 4.2. FCM is applied to classify these lots into five categories, based on their six input variables [6]. The required MATLAB code is provided in Fig. 4.4. Take example #1 as an example. It belongs to the five categories with memberships 0.070, 0.022, 0.884, 0.017, and 0.007, respectively. The example is classified into category $\arg\max_{c} \mu_{1(c)} = 3$. In addition, $T_1 = \max_{c} \mu_{1(c)} = 0.884$;

Table. 4.2 Collected data of 200 lots

j	x_{j1}	x_{j2}	x_{j3}	x_{j4}	x_{j5}	x_{j6}	y_j(hrs)
1	22	1320	165	748	421	85%	1298
2	23	1261	169	734	198	81%	1098
3	22	1187	159	635	287	93%	1122
…							
200	24	1309	175	734	345	81%	1276

```
cd F:\NCI
cycletimeforecastingexample=readtable("cycle time forecasting example.xlsx");

cycletimeforecastingexample=table2array(cycletimeforecastingexample);

train_x=cycletimeforecastingexample(1:150,2:7);
train_y=cycletimeforecastingexample(1:150,8:8);
test_x=cycletimeforecastingexample(151:200,2:7);
test_y=cycletimeforecastingexample(151:200,8:8);

[centers,U,objFcn]=fcm(train_x,5);

T1=max(U);
F1=1-T1;
I1=-5/4*T1+5/4;

figure;
plot([1:150],T1,'--o',[1:150],I1,'--k',[1:150],F1,'b');
title('Memberships');
xlabel('Example #');
ylabel('Membership');
legend('Truth membership','Indeterminacy membership','Falsity membership');
hold off;
```

Fig. 4.4 MATLAB code for implementing FCM

$F_1 = 1 - \max_c \mu_{1(c)} = 0.116$; $I_1 = \frac{-5}{5-1} \cdot 0.884 + \frac{5}{5-1} = 0.145$. Therefore, the classification result of example 1 is $y_1 = \langle 3; 0.884, 0.145, 0.116 \rangle$. The three memberships of all examples are summarized in Fig. 4.5.

4.4.3 Neutrosophic Classification Tree

Neutrosophic classification trees can be a supervised learning approach if hit rate is to be maximized during training. In a neutrosophic classification tree, the classification mechanism is composed of decision rules such as"

$$\mathrm{If}\, x_{jr_m(1)} \Omega s_m(1) \text{ AND } x_{jr_m(2)} \Omega s_m(2) \text{ AND} \ldots \text{THEN } o_j = G_m \quad (4.9)$$

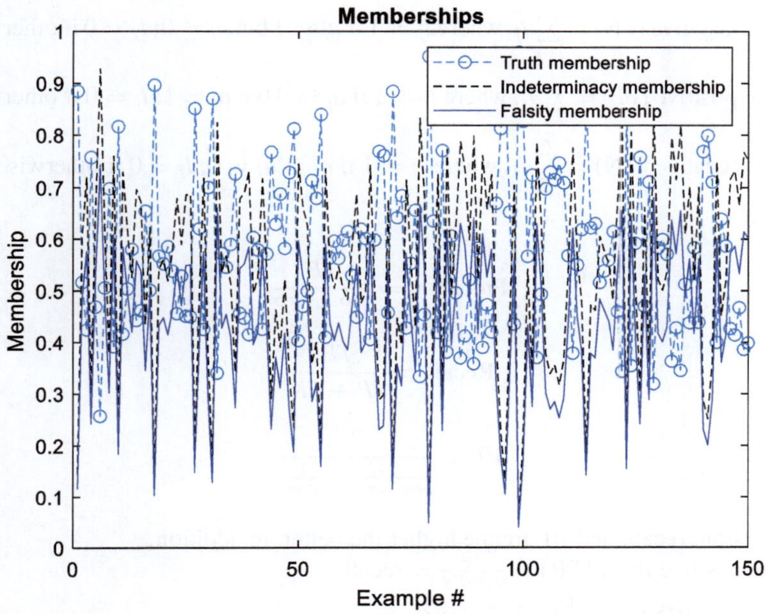

Fig. 4.5 Memberships of all examples in the training data

" where $\Omega \in \{<, \geq\}$; $m = 1 \sim M$. There are M decision rules in the classification tree. The classification results by the decision rules are aggregated using the softmax function as

$$o_j = \arg\max_{G_m} f(G_m) \tag{4.10}$$

where $f()$ is the frequency function.

The accuracy using the classification model can be evaluated in terms of hit rate as

$$HR = \frac{\sum_{j=1}^{n} I_j}{n} \tag{4.11}$$

where

$$I_j = \begin{cases} 1 & if \quad a_j = o_j \\ 0 & otherwise \end{cases} \tag{4.12}$$

where o_j and a_j indicate the classification result and actual classification of item j, respectively. In addition, for a binary classification task,

True positive (TP) $= \sum_{j=1}^{n} I_j$, where $I_j = 1$ if $a_j = o_j = 1$; $I_j = 0$ if otherwise.

False negative (FN) $= \sum_{j=1}^{n} I_j$, where $I_j = 1$ if $a_j = 1$ but $o_j = 0$; $I_j = 0$ if otherwise.

False positive (FP) $= \sum_{j=1}^{n} I_j$, where $I_j = 1$ if $a_j = 0$ but $o_j = 1$; $I_j = 0$ if otherwise.

True negative (TN) $= \sum_{j=1}^{n} I_j$, where $I_j = 1$ if $o_j = a_j = 0$; $I_j = 0$ if otherwise.

Then,

$$\text{Precision} = \frac{TP}{TP + FP}$$

$$\text{Recall} = \frac{TP}{TP + FN}$$

$$\text{F1} = \frac{2}{\frac{1}{precision} + \frac{1}{recall}}$$

Precision, recall, and F1 are the higher the better. In addition,
True positive rate (TPR) $= \frac{TP}{TP+FN} =$ recall.
False positive rate (FPR) $= \frac{FP}{FP+TN}$.
A receiver operation characteristic (ROC) curve can be plotted, in which the x-axis and the y-axis are FPR and TPR, respectively. The area under curve (AUC) is between [0, 1] and is also the higher the better.

Example 4.2. The diabetes dataset from the National Institute of Diabetes and Digestive and Kidney Diseases (NIDDKD) [8] is used to illustrate the applicability of a neutrosophic classification tree, in which whether a patient has diabetes is classified based on her attributes including.

- Number of pregnancies;
- Blood glucose level;
- Blood pressure measurement;
- Skin thickness;
- Blood insulin level;
- Body mass index (BMI);
- Diabetes pedigree function, which estimates the likelihood of a patient developing diabetes based on his/her family history;
- Age.

Instances in the dataset are selected from a large database subject to several limitations [8]. For example, all patients are females of Pima Indian heritage and were 21 years of age or older. A classification tree is built to this end following the fourfold cross-validation. The required MATLAB code is shown in Fig. 4.6. The classification accuracy is evaluated as $HR = 74.4\%$ (for the training data) and 74.1% (for test data). The trained classification tree is shown in Fig. 4.7. For example, the first classification rule is.

```
cd F:\NCI
training_data_table=readtable("diabetes-example.xlsx");
training_data=table2array(training_data_table);

n1=768;
cv1=4;
train_hit=zeros(cv1,1);
test_hit=zeros(cv1,1);
train_hitrate=zeros(cv1,1);
test_hitrate=zeros(cv1,1);

for i=1:cv1
   train_x=[training_data(1:round(n1*(1-i/cv1)),1:8);   training_data(round(n1*(1-(i-1)/cv1))+1:n1,1:8)];
   train_y=[training_data(1:round(n1*(1-i/cv1)),9:9);   training_data(round(n1*(1-(i-1)/cv1))+1:n1,9:9)];
   test_x=training_data(round(n1*(1-i/cv1))+1:round(n1*(1-(i-1)/cv1)),1:8);
   test_y=training_data(round(n1*(1-i/cv1))+1:round(n1*(1-(i-1)/cv1)),9:9);

   CT1=fitctree(train_x,train_y)
   train_o=CT1.predict(train_x);
   test_o=CT1.predict(test_x);

   train_hit(i,1)=0;
   test_hit(i,1)=0;
   for j=1:max(size(train_y))
      if train_o(j,1)==train_y(j,1)
         train_hit(i,1)=train_hit(i,1)+1;
      end
   end
   train_hitrate=train_hit/round(n1*(cv1-1)/cv1)

   for j=1:max(size(test_y))
      if test_o(j,1)==test_y(j,1)
         test_hit(i,1)=test_hit(i,1)+1;
      end
   end
   test_hitrate(i,1)=test_hit(i,1)/round(n1/cv1)

end

mean(train_hitrate)
mean(test_hitrate)
view(CT1,'Mode','graph');
```

Fig. 4.6 MATLAB code for implementing the classification tree for classifying patients

"If $x_{j6} < 26.35$ AND $x_{j2} < 138$ AND $x_{j1} < 9.5$ AND $x_{j7} < 0.7055$ THEN $o_j = 0$"

Namely,

"If BMI < 26.35 AND blood glucose level < 138 AND number of pregnancies < 9.5 AND diabetes pedigree function < 0.7055 THEN the patient does not have type-II diabetes"

The maximum number of splits in the classification tree in Fig. 4.7 is $n - 1$, where n is the training data size. The minimum leaf size is 1, i.e., each node should

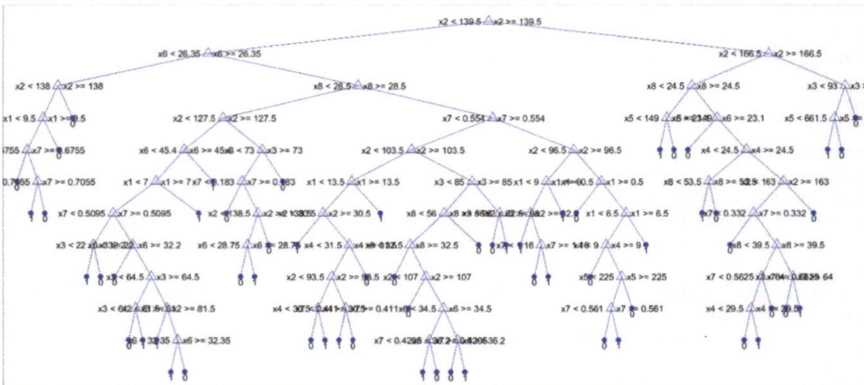

Fig. 4.7 Fitted classification tree

contain at least one example. The minimum parent size is 10, i.e., a node is split only if it contains at least 10 examples. These settings help control the depth of the classification tree.

Example 4.3. For the previous example, a ROC curve is plotted in Fig. 4.8. The AUC is 0.7515. The required MATLAB code is provided in Fig. 4.9.

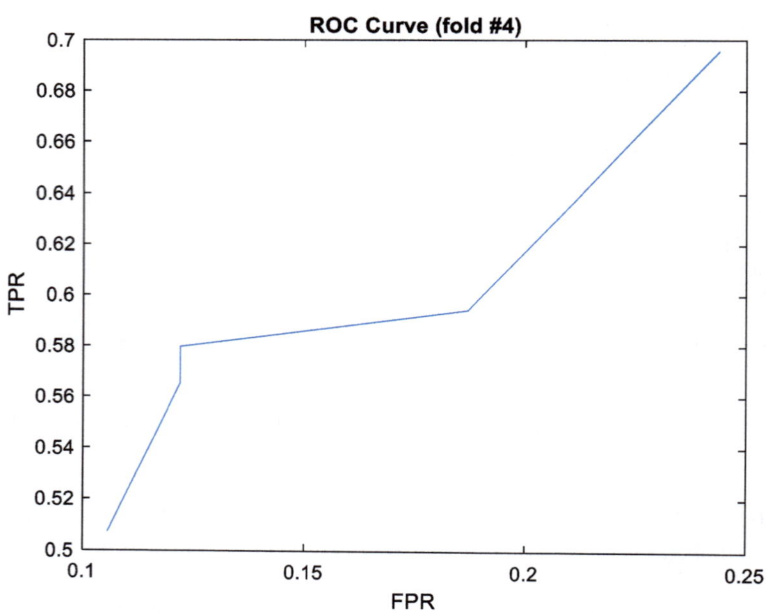

Fig. 4.8 ROC curve for Example 4.1

```
plot([0.1057 0.122 0.122 0.187 0.2439],[0.5072 0.5652 0.5797 0.5942 0.6957]);
trapz([0 0.1057 0.122 0.122 0.187 0.2439 1],[0 0.5072 0.5652 0.5797 0.5942 0.6957 1])
xlabel('FPR');
ylabel('TPR');
title('ROC Curve (fold #4)');
```

Fig. 4.9 MATLAB code for plotting the ROC curve

Since HR is the higher the better, the memberships of a neutrosophic classification result can be derived as

$$\alpha_{\tilde{o}_j} = e^{-\rho_1 \cdot (1-HR)} \tag{4.13}$$

$$\theta_{\tilde{o}_j} = 1 - e^{-\rho_2 \cdot \frac{(1-HR(prox(\mathbf{x_j})))}{|prox(\mathbf{x_j})|}} \tag{4.14}$$

$$\beta_{\tilde{o}_j} = 1 - e^{-\rho_1 \cdot (1-HR)} \tag{4.15}$$

in which $\rho_1, \rho_2 \in \mathbf{R}^+$, and

$$HR(prox(\mathbf{x_j})) = \frac{\sum_{i=1}^{n} I_i(prox(\mathbf{x_j}))}{n} \tag{4.16}$$

$prox(\mathbf{x_j})$ measures the neighborhood crowding degree of $\mathbf{x_j}$, reflecting the local learning sufficiency. $\mathbf{x_i} \in prox(\mathbf{x_j})$ if $dist(\mathbf{x_i}, \mathbf{x_j}) \leq \epsilon$, as illustrated in Fig. 4.10. The distance between two examples is measured using their Euclidean distance as

$$dist(\mathbf{x_i}, \mathbf{x_j}) = \sqrt{\sum_q (x_{iq} - x_{jq})^2} \tag{4.17}$$

Property 4.1. $\alpha_{\tilde{o}_j} \propto HR$.

Property 4.2. $\beta_{\tilde{o}_j} \propto \frac{1}{HR}$.

Property 4.3.

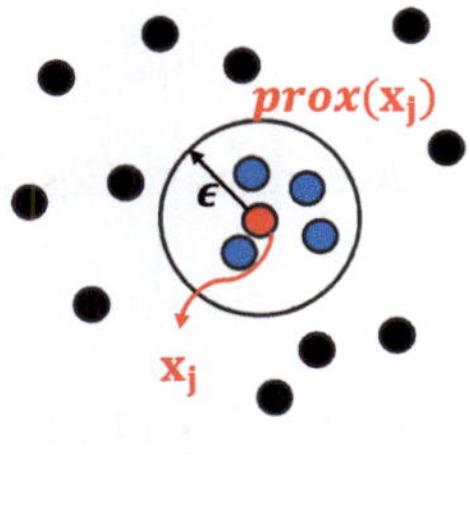

Fig. 4.10 Proximity of an example

- $\theta_{\tilde{o}_j} \propto \frac{1}{HR(prox(\mathbf{x_j}))}$
- $\theta_{\tilde{o}_j} \propto \frac{1}{|prox(\mathbf{x_j})|}$

The rationale is explained as follows. The accuracy of a classification model only indicates that its overall (average) classification performance has been optimized. Some examples still be incorrectly classified, especially those without neighboring data points, the proximity of which has been insufficiently learned. If the neighboring region of such examples is full of data points, it can be fully learned, and correct classifications are likely to be made. Another situation is that an example can still be correctly classified, even if the neighboring region of the example lacks examples. Therefore, both the local learning sufficiency (i.e., neighborhood crowding degree) and the local classification accuracy are considered in evaluating the indeterminacy membership.

Theorem 4.2 [9]

- $\alpha_{\tilde{o}_j}, \theta_{\tilde{o}_j}, \beta_{\tilde{o}_j} \in]-0, 1+[$
- $0^- \leq \alpha_{\tilde{o}_j} + \theta_{\tilde{o}_j} + \beta_{\tilde{o}_j} \leq 3^+$

Proof:

- $\rho_1 \geq 0$ and $1 - HR \geq 0$. So,

$$0 < \alpha_{\tilde{o}_j} = e^{-\rho_1 \cdot (1-HR)} \leq 1 \tag{4.18}$$

and

$$0 \leq \beta_{\tilde{o}_j} = 1 - \alpha_{\tilde{o}_j} < 1 \tag{4.19}$$

In addition, ρ_2, $1 - HR(prox(\mathbf{x_j}))$, and $|prox(\mathbf{x_j})| \geq 0$. As a result,

$$0 < e^{-\rho_2 \cdot \frac{(1-HR(prox(\mathbf{x_j})))}{|prox(\mathbf{x_j})|}} \leq 1 \tag{4.20}$$

Therefore,

$$0 \leq \theta_{\tilde{o}_j} = 1 - e^{-\rho_2 \cdot \frac{(1-HR(prox(\mathbf{x_j})))}{|prox(\mathbf{x_j})|}} \leq 1 \tag{4.21}$$

According to Eqs. (4.10) and (4.11),

$$\alpha_{\tilde{o}_j} + \beta_{\tilde{o}_j} = 1 \tag{4.22}$$

Adding Eq. (4.14) and Constraint (4.13) gives

$$0 \leq \alpha_{\tilde{o}_j} + \theta_{\tilde{o}_j} + \beta_{\tilde{o}_j} \leq 2 \tag{4.23}$$

Theorem 4.2 is proved.

If $\beta_{\tilde{o}_j} \geq \psi$ (a threshold), the classification result for the example is not trustworthy and should not be casually applied.

Example 4.4. In the previous example, the memberships of the classification result are derived. First, set ρ_1 to 0.4. Since *HR* for the training data is 74.4%,

$$\alpha_{\tilde{o}_j} = e^{-0.4\cdot(1-0.744)} = 0.903$$

Therefore,

$$\beta_{\tilde{o}_j} = 1 - \alpha_{\tilde{o}_j} = 0.097$$

Take the third patient in test data as an example. $\tilde{o}_j = 1$. Set $\in$ to 20. Four examples are in the proximity of this patient. The local classification accuracy around this patient is 75%. By setting ρ_2 to 5,

$$\theta_{\tilde{o}_j} = 1 - e^{-5\cdot\frac{(1-75\%)}{4}} = 0.268$$

Therefore, the neutrosophic classification result for this patient is $\langle 1; 0.903, 0.268, 0.097 \rangle$. By setting ψ to 0.5, obviously the classification result of this patient is trustworthy.

4.4.4 Optimizing the Values of Parameters

The effects of ρ_1 on the truth membership ($\alpha_{\tilde{o}_k}$) and the falsity memberships ($\beta_{\tilde{o}_k}$) are shown in Figs. 4.11 and 4.12, respectively. Obviously, as the value of ρ_1 increases, the truth membership decreases rapidly. The falsity membership is the opposite.

In addition, the effect of the local learning sufficiency (i.e., neighbor crowding degree) ($|prox(\mathbf{x_j})|$) on the indeterminacy membership ($\theta_{\tilde{o}_j}$) is shown in Fig. 4.13. As expected, as the local learning sufficiency increases, the indeterminacy membership decreases. This means that with more learning from similar examples, users are more determined in accepting, trusting, and communicating classification results. However, when local learning accuracy gets lower, the effect of the local learning sufficiency becomes less obvious, meaning that the classification model repeatedly makes mistakes in classifying examples.

The values of the two real-valued parameters ρ_1 and ρ_2 can be set to meet specific user expectations [10]. For example, to restrict the value of $\alpha_{\tilde{o}_j}$ to be within [0.01, 0.99], i.e., to avoid extremely high or low truth memberships, it is easy to derive

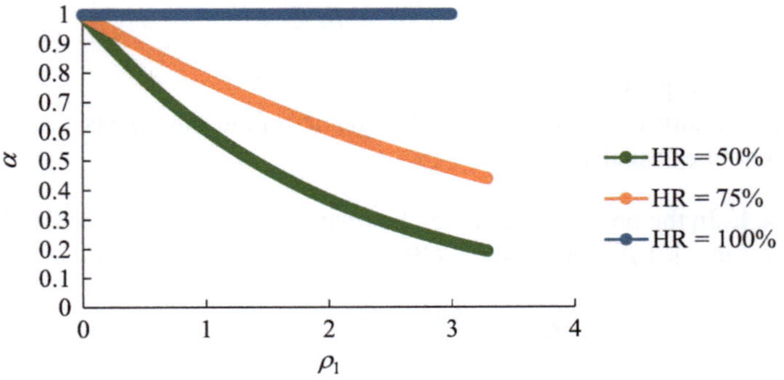

Fig. 4.11 Effect of ρ_1 on the truth membership ($\alpha_{\tilde{o}_j}$)

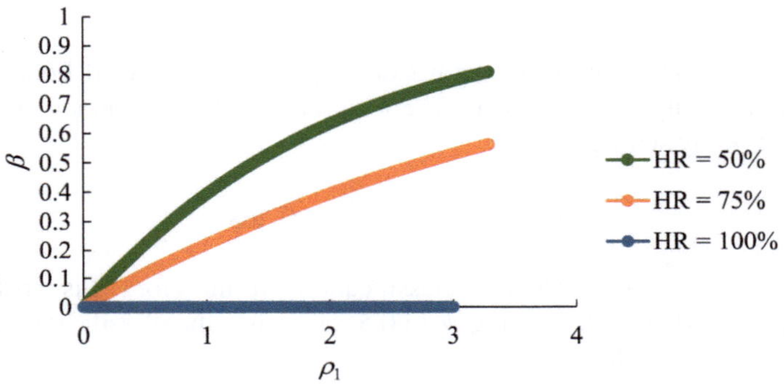

Fig. 4.12 Effect of ρ_1 on the falsity memberships ($\beta_{\tilde{o}_j}$)

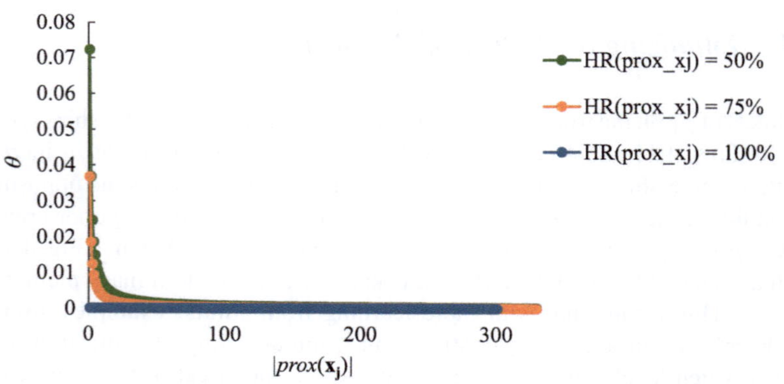

Fig. 4.13 Effect of local learning sufficiency on the indeterminacy membership

Table. 4.3 Possible results of deriving the value of ρ_1

HR	Range of ρ_1
100%	[101, 46052]
75%	[0.04, 18.413]
50%	[0.02, 9.208]
25%	[0.013, 6.139]
0%	[0.01, 4.605]

Table. 4.4 Possible results of deriving the value of ρ_2

$HR\left(prox(\mathbf{x_j})\right)$	Range of ρ_2
100%	[503, 230259]
75%	[0.201, 92.067]
50%	[0.1, 46.042]
25%	[0.067, 30.697]
0%	[0.05, 23.024]

$$\frac{-\ln 0.99}{(1-HR)} \leq \rho_1 \leq \frac{-\ln 0.01}{(1-HR)} \tag{4.24}$$

Possible results are summarized in Table 4.3. Similarly, to avoid the falsity membership from getting very high or low, the value of ρ_2 can be chosen from the range,

$$\frac{-\ln 0.99}{\frac{(1-HR(prox(\mathbf{x_j}))}{|prox(\mathbf{x_j})|}} \leq \rho_2 \leq \frac{-\ln 0.01}{\frac{(1-HR(prox(\mathbf{x_j}))}{|prox(\mathbf{x_j})|}} \tag{4.25}$$

Possible results when $|prox(\mathbf{x_j})| = 5$, i.e., there are five examples around an example, are summarized in Table 4.4.

In practice, if $HR \geq 95\%$, to ensure that $\alpha_{\tilde{o}_j} \geq 0.99$,

$$e^{-\rho_1\cdot(1-0.95)} \geq 0.99$$

$$\rho_1 \leq \frac{\ln(0.99)}{-0.05} = 0.201$$

In contrast, if $HR \leq 5\%$ (hrs), to ensure that $\alpha_{\tilde{o}_k} \leq 0.01$,

$$e^{-\rho_1\cdot(1-0.05)} \leq 0.01$$

$$\rho_1 \geq \frac{\ln(0.01)}{-0.95} = 4.848$$

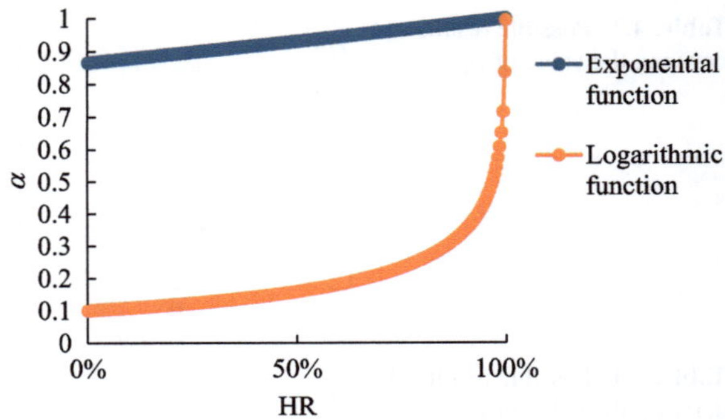

Fig. 4.14 Comparison of different functions for deriving memberships ($\rho_1 = 0.144$)

Furthermore, in the formulas for deriving the memberships, exponential functions are employed. Theoretically, other types of functions are also applicable, such as logarithmic functions:

$$\alpha_{\tilde{o}_j} = \rho_1 \ln\left(\frac{1}{(1 - HR)} + 1\right) \tag{4.26}$$

$$\theta_{\tilde{o}_j} = 1 - \rho_2 \ln\left(\frac{1}{\frac{(1-HR(prox(\mathbf{x_j}))}{|prox(\mathbf{x_j})|}} + 1\right) \tag{4.27}$$

$$\beta_{\tilde{o}_k} = 1 - \rho_1 \ln\left(\frac{1}{(1 - HR)} + 1\right) \tag{4.28}$$

The effects of the two types of functions on the derived truth memberships are compared in Fig. 4.14. Obviously, the logarithmic function is more sensitive to the decline in the classification accuracy in terms of HR.

4.5 Collaboration

Multiple neutrosophic classification methods can be simultaneously applied to classify an item. Then, a neutrosophic collaborative classification approach aggregates the classification results by several neutrosophic classification methods [11]. Let $\tilde{o}_j(m) = \langle o_j(m); \alpha_{\tilde{o}_j(m)}, \theta_{\tilde{o}_j(m)}, \beta_{\tilde{o}_j(m)} \rangle$ be the classification result of item j by neutrosophic classification method m. Then, the classification results are aggregated using softmax as

$$\tilde{o}_j = softmax(\{\tilde{o}_j(m)\}) = \langle o_j; \alpha_{\tilde{o}_j}, \theta_{\tilde{o}_j}, \beta_{\tilde{o}_j} \rangle \tag{4.29}$$

where

Table. 4.5 Classification of a patient by three neutrosophic classification methods

m	$\tilde{o}_j(m)$
1	<1; 0.85, 0.35, 0.15>
2	<0; 0.76, 0.42, 0.34>
3	<1; 0.90, 0.43, 0.10>

$$o_j = \arg\max_{o_j(m)} f\big(o_j(m)\big) \tag{4.30}$$

$$\alpha_{\tilde{o}_j} = \min_{o_j(m)=o_j} \alpha_{\tilde{o}_j(m)} \tag{4.31}$$

$$\theta_{\tilde{o}_j} = \max_{o_j(m)=o_j} \theta_{\tilde{o}_j(m)} \tag{4.32}$$

$$\beta_{\tilde{o}_j} = \max_{o_j(m)=o_j} \beta_{\tilde{o}_j(m)} \tag{4.33}$$

Example 4.5. Three neutrosophic classification methods are applied to classify a patient. The classification results are summarized in Table 4.5. Therefore,

$f\big(o_j(m)\big) = 1$ when $o_j(m) = 0$.
$f\big(o_j(m)\big) = 2$ when $o_j(m) = 1$

As a result,

$$o_j = \arg\max_{o_j(m)} f\big(o_j(m)\big) = 1$$

$$\alpha_{\tilde{o}_j} = \min_{o_j(m)=1} \alpha_{\tilde{o}_j(m)} = \min(0.85, 0.90) = 0.85$$

$$\theta_{\tilde{o}_j} = \max_{o_j(m)=1} \theta_{\tilde{o}_j(m)} = \max(0.35, 0.43) = 0.43$$

$$\beta_{\tilde{o}_j} = \max_{o_j(m)=1} \beta_{\tilde{o}_j(m)} = \max(0.15, 0.10) = 0.15$$

Finally, $\tilde{o}_j = 1; 0.85, 0.43, 0.15$.

4.6 Neutrosophic Deep Neural Networks

A neutrosophic prediction method is also applicable to classification problems. The prediction result is mapped to one of several categories. Possibility prediction results [12] can provide a better basis for collaboration between neutrosophic classification methods.

For example, a deep neural network (DNN) can be constructed and trained to predict the target variable used in classification based on the input attributes. The DNN is a feed-forward, fully connected neural network [13]. The configuration of the DNN is as follows:

- Input layer: For each example, there are Q inputs to the DNN, $\{x_{jq}|q = 1 \sim Q\}$, indicating the attributes of example j; $j = 1 \sim n$.
- Hidden layers: The DNN has H hidden layers. The activation function on the nodes of the hidden layers is hyperbolic tangent (tanh). In the first hidden layer,

$$h_{jr_1}^{(1)} = \frac{2}{1 + e^{-2n_{jr_1}^{(1)}}} - 1 \tag{4.34}$$

where $h_{jr_1}^{(1)}$ is the output from the r_1-th node for example j,

$$n_{jr_1}^{(1)} = I_{jr_1}^{(1)} - \theta_{r_1}^{(1)} \tag{4.35}$$

$$I_{jr_1}^{(1)} = \sum_{q=1}^{Q} w_{qr_1} x_{jq} \tag{4.36}$$

where w_{qr_1} is the connection weight between the q-th input node and the r_1-th node of the first hidden layer; $\theta_{r_h}^{(h)}$ is the threshold on the r_h-th node of the h-th hidden layer. From the second hidden layer onwards,

$$h_{jr_h}^{(h)} = \frac{2}{1 + e^{-2n_{jr_h}^{(h)}}} - 1 \tag{4.37}$$

where

$$n_{jr_h}^{(h)} = I_{jr_h}^{(h)} - \theta_{r_h}^{(h)} \tag{4.38}$$

$$I_{jr_h}^{(h)} = \sum_{r_{h-1}=1}^{R_{h-1}} w_{r_{h-1}r_h}^{(h-1)} h_{jr_{h-1}}^{(h-1)} \tag{4.39}$$

- Output (o_j): The value of the target variable of example j for classification. To this end, the linear transfer function (purelin) is adopted:

$$o_j = \sum_{r_H=1}^{R_H} w_{r_H}^{(o)} h_{jr_H}^{(H)} - \theta^o \tag{4.40}$$

$w_{r_H}^{(o)}$ is the connection weight; θ^o is the threshold of the output node. Therefore, o_j can be negative or greater than 1, covering both actual values.

- Training algorithm: Levenberg–Marquardt (LM) algorithm [14]
- Performance measure: root mean squared error (RMSE),

$$RMSE = \sqrt{\frac{\sum_{j=1}^{n}(o_j - a_j)^2}{n}} \tag{4.41}$$

The neutrosophic forecast of example j is

$$\tilde{o}_j = o_j; \alpha_{\tilde{o}_j}, \theta_{\tilde{o}_j}, \beta_{\tilde{o}_j} \tag{4.42}$$

In addition, based on the prediction results, the classification performance is evaluated as

$$HR = \frac{\sum_{j=1}^{n} I_j}{n} \tag{4.43}$$

For a binary classification problem,

$$I_j = \begin{cases} 1 \quad if \quad a_j = 1 \text{and} o_j \geq \xi \\ 1 \quad if \quad a_j = 0 \text{and} o_j < \xi \\ 0 \; otherwise \end{cases} \tag{4.44}$$

In Eqs. (4.42), the memberships of $\tilde{o}_j$ can be derived based on *HR* using Eqs. (4.26)–(4.28), or based on *RMSE* instead:

$$\alpha_{\tilde{o}_j} = \rho_3 \ln\left(\frac{1}{RMSE} + 1\right) \tag{4.45}$$

$$\theta_{\tilde{o}_j} = 1 - \rho_4 \ln\left(\frac{1}{\frac{RMSE(prox(\mathbf{x_j})}{|prox(\mathbf{x_j})|}} + 1\right) \tag{4.46}$$

$$\beta_{\tilde{o}_j} = 1 - (\rho_3 \ln\left(\frac{1}{RMSE} + 1\right)) \tag{4.47}$$

Example 4.6. For Example 4.2, a deep neural network (DNN) is constructed and trained in this study to diagnose whether these patients had diabetes based on their eight attributes. Four-fold cross-validation was followed. Therefore, the DNN is trained using 576 instances each fold, and then is applied to the remaining patients in test data. The confusion plot for test data (all folds combined; row normalization) is shown in Fig. 4.15.

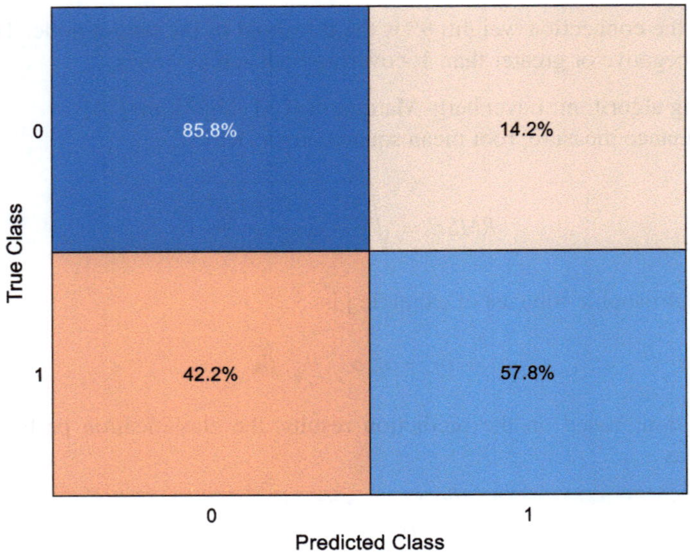

Fig. 4.15 Confusion plot for test data (all folds combined; row normalization)

Set $\in$, ρ_1, and ρ_2 to 20, 0.5, and 5, respectively. Consider fold #4. $HR = 77.4\%$. Therefore,

$$\alpha_{\tilde{o}_j} = e^{-\rho_1 \cdot (1-HR)} = e^{-0.5 \cdot (1-0.774)} = 0.893$$

$$\beta_{\tilde{o}_j} = 1 - 0.893 = 0.107$$

Consider the first patient in test data, patient #1. Three examples in the training data are in the proximity of this patient, so $|prox(\mathbf{x}_1)| = 3$. The local classification accuracy in terms of $HR(prox(\mathbf{x}_1)) = 67\%$. Therefore,

$$\theta_{\tilde{o}_1} = 1 - e^{-5 \cdot \frac{(1-0.67)}{3}} = 0.423$$

This patient is diagnosed as having type-II diabetes. Therefore, the classification result is 1; 0.893, 0.423, 0.107.

The classification results using multiple neutrosophic deep neural networks can be aggregated using the collaboration mechanism introduced in Sect. 4.3.

Example 4.7. Chen and Lin [5] proposed a neutrosophic classification method for trustworthy predictive maintenance in smart manufacturing, in which a DNN is constructed and trained to predict the next time to failure of a machine (o_j) based on three sensor readings for monitoring vibration (acceleration), temperature, and frequency, so as to arrange preventive maintenance in advance. The prediction results are summarized in Fig. 4.16. $RMSE = 3.43$ (hr).

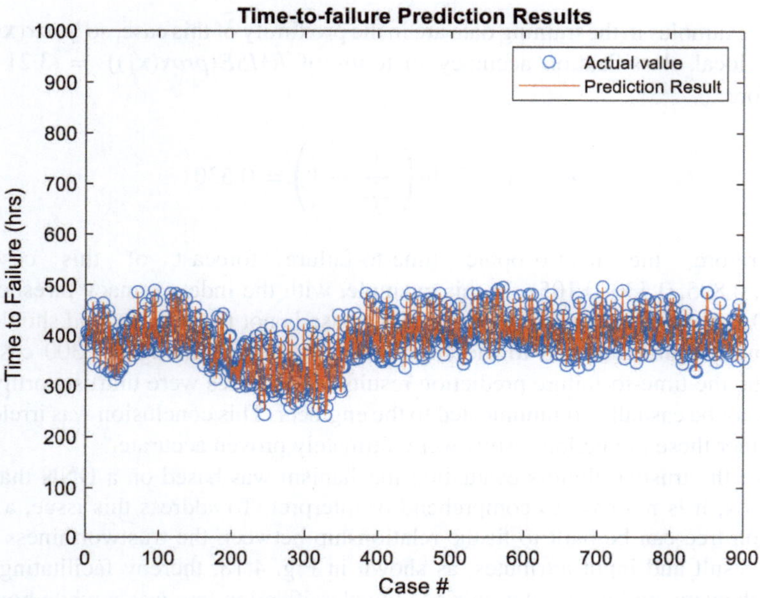

Fig. 4.16 Time-to-failure prediction results [5]

If $\theta_{o_j} \geq \psi$ (a threshold), the neutrosophic time-to-failure forecast was not trustworthy and should not be casually communicated to the engineer. Otherwise, if the predicted time to failure was earlier than a target value, a preventive maintenance should be arranged, as illustrated in Fig. 4.17. Consider the first case in test data for fold #4. $o_1 = 401.8$ (hr). Set $\in$, ρ_3, and ρ_4 to 20, 3.5, and 0.5, respectively:

$$\alpha_{\tilde{o}_1} = 3.5 \cdot \ln\left(\frac{1}{3.43} + 1\right) = 0.895$$

$$\beta_{\tilde{o}_1} = 1 - 0.895 = 0.105$$

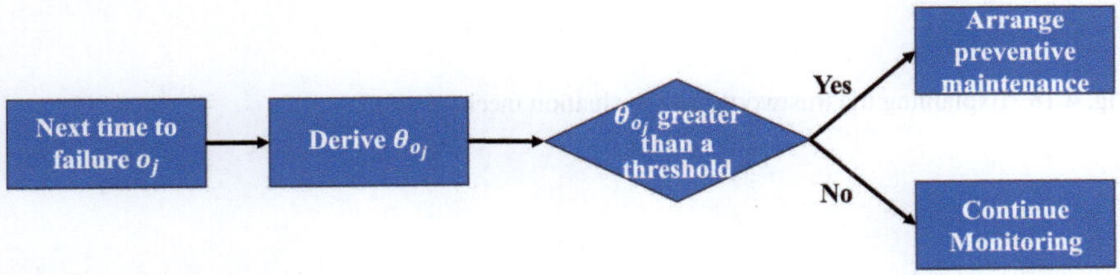

Fig. 4.17 Applications of neutrosophic classification for predictive maintenance

Five examples in the training data are in the proximity of this case, so $|prox(\mathbf{x}_1)| = 5$. The local classification accuracy in terms of $RMSE(prox(\mathbf{x}_1)) = 3.21$ (hr). Therefore,

$$\theta_{\tilde{o}_1} = 1 - 5 \cdot \ln\left(\frac{1}{\frac{3.21}{5}} + 1\right) = 0.530$$

Therefore, the neutrosophic time-to-failure forecast of this case is $\langle 401.8; 0.895, 0.530, 0.105 \rangle$. In this example, with the indeterminacy threshold ψ set at 0.5, the time-to-failure forecast of the case is not trustworthy and should not be casually communicated to the engineer. In addition, among the 300 cases in test data, the time-to-failure prediction results for 17 cases were untrustworthy and should not be casually communicated to the engineer. This conclusion was irrelevant to whether these prediction results were ultimately proven accurate.

Since the trustworthiness evaluation mechanism was based on a DNN that is a black box, it is not easy to comprehend or interpret. To address this issue, a classification tree can be built to fit the relationship between the trustworthiness evaluation result and input attributes, as shown in Fig. 4.18, thereby facilitating user comprehension and interpretation since the classification tree was a white box [15, 16].

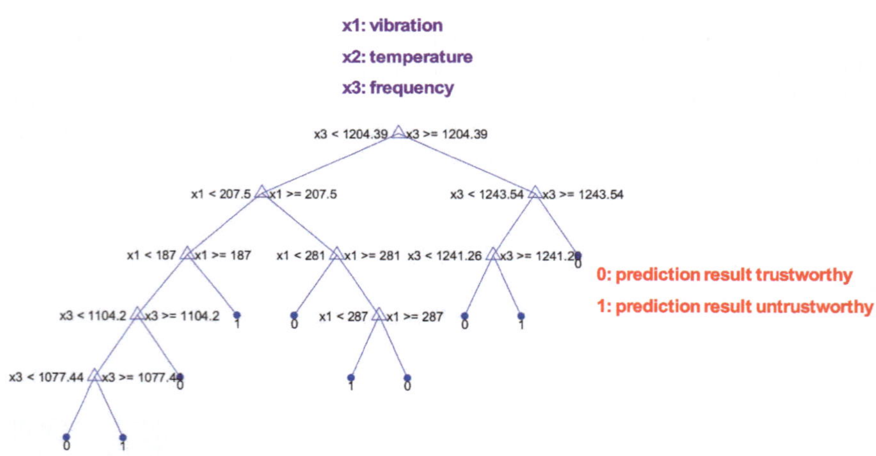

Fig. 4.18 Explaining the trustworthiness evaluation mechanism [5]

4.7 Predictive Collaboration

The prediction result using a neutrosophic prediction methods can be compared with a pre-specified threshold to classify the prediction result into one of several categories:

$$\tilde{c}_j(m) = \left\langle r; \alpha_{\tilde{o}_j(m)}, \theta_{\tilde{o}_j(m)}, \beta_{\tilde{o}_j(m)} \right\rangle \text{ if } L_r \leq o_j(m) < U_r \tag{4.48}$$

In addition, then, the classification results using several such neutrosophic predictive classification methods can be aggregated using softmax as

$$\tilde{c}_j = softmax(\{\tilde{c}_j(m)\}) = c_j; \alpha_{\tilde{c}_j}, \theta_{\tilde{c}_j}, \beta_{\tilde{c}_j} \tag{4.49}$$

where

$$c_j = \arg\max_{c_j(m)} f\left(c_j(m)\right) \tag{4.50}$$

$$\alpha_{\tilde{c}_j} = \min_{c_j(m)=c_j} \alpha_{\tilde{c}_j(m)} \tag{4.51}$$

$$\theta_{\tilde{c}_j} = \max_{c_j(m)=c_j} \theta_{\tilde{c}_j(m)} \tag{4.52}$$

$$\beta_{\tilde{c}_j} = \max_{c_j(m)=c_j} \beta_{\tilde{c}_j(m)} \tag{4.53}$$

Example 4.8. Three neutrosophic classification methods are applied to classify the future status of a machine by predicting its time to next failure. The prediction results are summarized in Table 4.6. Based on the prediction results, the classification results of the machine within the time interval [300, 400] (hr) are made in Table 4.7. The classification results using the three neutrosophic classification methods are aggregated as.

$$c_j = \arg\max_{c_j(m)} f\left(c_j(m)\right) = 1$$

$$\alpha_{\tilde{c}_j} = \min_{c_j(m)=1} \alpha_{\tilde{c}_j(m)} = \min(0.79, 0.81) = 0.81$$

Table. 4.6 Prediction results of the time to next failure of a machine

m	$o_j(m)$
1	<306; 0.79, 0.45, 0.21>
2	<294; 0.85, 0.24, 0.15>
3	<385; 0.81, 0.47, 0.19>

Table. 4.7 Classification results of the future status of the machine

m	$c_j(m)$
1	<1; 0.79, 0.45, 0.21>
2	<0; 0.85, 0.24, 0.15>
3	<1; 0.81, 0.47, 0.19>

$$\theta_{\tilde{c}_j} = \max_{c_j(m)=1} \theta_{\tilde{c}_j(m)} = \max(0.45, 0.47) = 0.47$$

$$\beta_{\tilde{c}_j} = \max_{c_j(m)=1} \beta_{\tilde{c}_j(m)} = \max(0.21, 0.19) = 0.21$$

Finally, $\tilde{c}_j = 1; 0.81, 0.47, 0.21$. Therefore, the machine is likely to fail during the time interval with a possibility of 0.81. Such a classification result is trustworthy since $\theta_{\tilde{c}_j} = 0.47 < 0.5$ (the pre-specified threshold).

References

1. G. Dhingra, V. Kumar, H.D. Joshi, A novel computer vision based neutrosophic approach for leaf disease identification and classification. Measurement **135**, 782–794 (2019)
2. A.Q. Ansari, R. Biswas, S. Aggarwal, Neutrosophic classifier: An extension of fuzzy classifer. Appl. Soft Comput. **13**(1), 563–573 (2013)
3. I. Kandasamy, F. Smarandache, Triple refined indeterminate neutrosophic sets for personality classification. in *2016 IEEE Symposium Series on Computational Intelligence* (2016), pp. 1–8
4. M. Sharma, I. Kandasamy, W.B. Vasantha, Emotion quantification and classification using the neutrosophic approach to deep learning. Appl. Soft Comput. **148**, 110896 (2023)
5. T.C.T. Chen, C.W. Lin, Trustworthy predictive maintenance in smart manufacturing using neutrosophic reasoning approach. Inter. J. Adv. Manuf. Tech. 1–13 (2026)
6. T. Chen, Y.C. Wang, A two-stage explainable artificial intelligence approach for classification-based job cycle time prediction. Int. J. Adv. Manuf. Technol. **123**, 2031–2042 (2022)
7. A. Singh, H. Kulkarni, F. Smarandache, G.K. Vishwakarma, Computation of separate ratio and regression estimator under neutrosophic stratified sampling: an application to climate data. J. Fuzzy Ext. Appl. **5**(4), 605–621 (2024)
8. A.D. Khare, Diabetes dataset (2022). https://www.kaggle.com/datasets/akshaydattatraykhare/diabetes-dataset?resource=download
9. N. El-Hefenawy, M.A. Metwally, Z.M. Ahmed, I.M. El-Henawy, A review on the applications of neutrosophic sets. J. Comput. Theor. Nanosci. **13**(1), 936–944 (2016)
10. A. Kordi, A. Nemati, Simultaneous sensitivity analysis of mixed-integer location-allocation models using machine learning tools: Cancer hospitals' network design. Oper. Res. Int. Journal **24**(2), 22 (2024)
11. T. Chen, C.-W. Lin, A fuzzy collaboration system for ubiquitous loading/unloading space recommendation in the logistics industry. Robot. Comput.-Integr. Manuf. **45**, 86–98 (2017)
12. M. Kusy, J. Kluska, Assessment of prediction ability for reduced probabilistic neural network in data classification problems. Soft. Comput. **21**(1), 199–212 (2017)
13. R.S. Tang, Y.M. Li, Y. Bao, J.D. Zeng, Y.J. Yao, J.Y. Wang, M. Chen, Development of a multi-label deep neural network model for predicting immediate paravalvular leakage and new-onset conduction disturbances after transcatheter aortic valve replacement: a retrospective cohort study. Digital Health **12**, 20552076261427502 (2026)

14. H.P. Gavin, The Levenberg-Marquardt algorithm for nonlinear least squares curve-fitting problems. Department of Civil and Environmental Engineering Duke University August 3(1–23), 4 (2019)
15. Y.C. Wang, T. Chen, Adapted techniques of explainable artificial intelligence for explaining genetic algorithms on the example of job scheduling. Expert Syst. Appl. **237**, 121369 (2024)
16. T.C.T. Chen, C.W. Lin, Y.C. Wang, Visualization XAI techniques for facilitating schedulers' comprehension and interpretation of GA applications in job scheduling. Int. J. Hum.-Comput. Interact. **42**(1), 638–652 (2026)
17. T.C.T. Chen, Y.C. Wang, Y.C. Wang, An explainable decision model for selecting facility locations in supply chain networks. Supply Chain. Anal. **11**, 100148 (2025)

MIX
Papier aus verantwortungsvollen Quellen
Paper from responsible sources
FSC® C105338

If you have any concerns about our products,
you can contact us on
ProductSafety@springernature.com

In case Publisher is established outside the EU,
the EU authorized representative is:
Springer Nature Customer Service Center GmbH
Europaplatz 3, 69115 Heidelberg, Germany

Printed by Libri Plureos GmbH
in Hamburg, Germany